Representing Landscapes

This volume provides an in-depth historical overview of graphic and visual communication styles, techniques, and outputs from key landscape architects over the past century. *Representing Landscapes: One Hundred Years of Visual Communication* offers a detailed account of how past and present landscape architects and practitioners have harnessed the power of visualization to frame and situate their designs within the larger cultural, social, ecological, and political milieux.

The fifth book in the Representing Landscapes series, the presentations contained within each of the 25 chapters of this work are not merely drawings and illustrations but are rather graphic touchstones whose past and current influence shapes how landscape architects think and operate within the profession. This collected volume of essays gathers notable landscape historians, scholars, and designers to offer their insights on how the landscape has been presented and charts the development and use of new technologies and contemporary theory to reveal the conceptual power of the living medium of the larger landscape.

Richly detailed with over 220 colour and black and white illustrations from some of the discipline's best-known landscape architects and designers, this work is a 'must-have' for those studying contemporary landscape design or those fascinated by the profession's history.

Nadia Amoroso, PhD, OALA, CSLA, is an Associate Professor in Landscape Architecture at the University of Guelph, School of Environmental Design and Rural Development. She holds a PhD from the Bartlett School of Architecture, UCL, London, and degrees in Landscape Architecture and Urban Design from the University of Toronto. She specializes in visual communication in landscape architecture, digital design, data visualization, and creative mapping. She also operates an illustration studio, under her name, focusing on landscape architectural visual communication. She has written a number of articles and books on topics relating to creative mapping, visual representation, and digital design including *The Exposed City: Mapping the Urban Invisibles*, *Representing Landscapes: Digital*, *Representing Landscapes: Hybrid* and *Digital Landscape Architecture Now*.

Martin J. Holland, PhD, is an Assistant Professor in the School of Environmental Design and Rural Development at the University of Guelph, located in Southwestern Ontario, Canada. Dr. Holland teaches a range of courses and studios in landscape design, urban design, and landscape history and theory. He has taught studio courses at Clemson University, University of Illinois at Urbana-Champaign, and the Illinois Institute of Technology (IIT) in Chicago. His scholarly interests lie at the intersection of landscape design, cultural studies, and collective memory. He is particularly interested in how monuments, memorials, and other sites of commemoration are used, managed, and interpreted to guide, inform, and influence the public's understanding of history and how it relates to the built environment. Professor Holland received his doctorate from the University of Illinois at Urbana-Champaign and his MLA from the University of Virginia. He completed his bachelor's degree at Dalhousie University in Halifax, Nova Scotia, where he majored in philosophy.

Representing Landscapes

A Visual Collection of Landscape Architectural Drawings
Nadia Amoroso

Representing Landscapes
Digital
Nadia Amoroso

Representing Landscapes
Hybrid
Nadia Amoroso

Representing Landscapes
Analogue
Nadia Amoroso

Representing Landscapes
One Hundred Years of Visual Communication
Nadia Amoroso and Martin J. Holland

For more information about this series, please visit: https://www.routledge.com/Representing-Landscapes/book-series/REPLAND

Representing Landscapes

One Hundred Years of Visual Communication

Edited by
Nadia Amoroso and Martin J. Holland

NEW YORK AND LONDON

Cover credit: GROSS.MAX. / Ross Ballard; Kathryn Gustafson / GNN; Walter Hood; Garrett Eckbo; Chip Sullivan; James Corner.

First published 2022
by Routledge
605 Third Avenue, New York, NY 10158

and by Routledge
4 Park Square, Milton Park, Abingdon, Oxon, OX14 4RN

Routledge is an imprint of the Taylor & Francis Group, an informa business

© 2022 selection and editorial matter, Nadia Amoroso and Martin J. Holland; individual chapters, the contributors

The right of Nadia Amoroso and Martin J. Holland to be identified as the authors of the editorial material, and of the authors for their individual chapters, has been asserted in accordance with sections 77 and 78 of the Copyright, Designs and Patents Act 1988.

All rights reserved. No part of this book may be reprinted or reproduced or utilised in any form or by any electronic, mechanical, or other means, now known or hereafter invented, including photocopying and recording, or in any information storage or retrieval system, without permission in writing from the publishers.

Trademark notice: Product or corporate names may be trademarks or registered trademarks, and are used only for identification and explanation without intent to infringe.

Library of Congress Cataloging-in-Publication Data
A catalog record for this title has been requested

ISBN: 9781032024554 (hbk)
ISBN: 9781032024547 (pbk)
ISBN: 9781003183402 (ebk)

DOI: 10.4324/9781003183402

Typeset in Garamond
by codeMantra

Contents

Notes on Contributors · vii
Acknowledgements · xii

1. Introduction · 1
   Nadia Amoroso

2. A System of Expression: Writing and Making Landscapes of Gertrude Jekyll · 4
   Halina Steiner

3. Beatrix Farrand: Representing Landscape in Prose and Drawings · 13
   Thaisa Way

4. Fletcher Steele, the Savvy Practitioner: Desire and the Cultivation of Connoisseurship · 24
   Martin J. Holland

5. Topographical and Landform Explorations: Revisiting Noguchi's Sculptured Landscapes and Their Representations · 32
   Shannon Bassett

6. Burle Marx: The Individual Language of Plenitude · 43
   Ana Rita Sá Carneiro

7. J.B. Jackson: Representing Everyday Landscapes · 51
   Jeffrey D. Blankenship

8. The EDSA Style: "A Legacy of Graphic Communication" · 59
   Kona A. Gray

9. Boomerangs, Zig-Zags, and Orbits: Drawing the California Garden Garrett Eckbo and Thomas Church · 67
   Chip Sullivan

10. The Drawings of Lawrence Halprin · 77
    Alison Hirsch

11. Ian L. McHarg and Mapping Complex Processes · 86
    Frederick R. Steiner

12. Drawing Experiments for Representing Landscape · 99
    Noémie Lafaurie-Debany, Javier González-Campaña, and Balmori Associates

13. Peter Walker: The Growth of Representation · 107
    Peter Walker

14. Pieces of the World: Yves Brunier's Landscape Representations · 116
    Linda Pollak

15. Hands On! · 124
    Petra Blaisse

16. Freedom from an Innocent Landscape: The Visual Communication of West 8 · 132
    Adriaan Geuze

17. Evolving Representation, Physical and Digital at Hargreaves Jones Landscape Architecture · 144
    Matt Perotto

18 Drawing *in* Perspective 154
    David Malda

19 The Eidetic Drawings of James Corner 161
    Tina George and Nadia Amoroso

20 Non-sites and Simulacra 170
    Ken Smith

21 The Spirit of Drawing 177
    Chip Sullivan

22 Allegorical Drawings: Developing a Cultural Practice 185
    Walter J. Hood

23 ASPECTS [of] Design Representation 192
    ASPECT Studios and Jillian Walliss

24 Every Picture Tells a Story: The Iconography of GROSS.MAX. Imagery 199
    Eelco Hooftman

Final Thoughts 209
Nadia Amoroso and Martin J. Holland

Index 213

Notes on Contributors

ASPECT Studios are a group of designers united through a philosophy that delivers innovative landscape architecture, placemaking, and urban design. The studios are connected by their purpose: to create projects which challenge and delight, whilst enhancing the lives of people and natural systems in an enduring way. With nine studios across Australia, Asia, and the Middle East, they draw on the expertise of a global team to locally deliver site responsive, generous places where people want to be.

Balmori Associates, a New York City-based urban and landscape design firm, is recognized worldwide for its creative interfacing of landscape and architecture. Through research, collaboration, and innovation, Balmori Associates explores and expands the boundaries between nature and structure. Our firm's approach is rooted in coupling design ideas with an understanding of environmental, social, and physical needs. Balmori Associates has established a signature aesthetic by applying inventive design thinking to a careful study of the social aspects of ecological, hydrological, and temporal dimensions of projects. As leaders in the field of urban design and innovative public space making, we use design to give form to the processes of sustainability. Whether developing new parks, redesigning existing open spaces, or creating recreational venues that integrate architecture and landscape, we seek to produce 'green infrastructures' with designs that invite community interaction while revealing the constructed and natural operations of a site.

Shannon Bassett is an Assistant Professor of Architecture at Laurentian University. She holds a Masters of Architecture in Urban Design from the Harvard Graduate School of Design and a Bachelors of Architecture with Distinction from Carleton University. She is a Canadian-American Architect and Urban Designer. Her research, teaching, writing, and practice operate at the intersection(s) of architecture, urban design, and landscape ecology. Her design work and research has been exhibited internationally, including at the Hong Kong Shenzhen Bi-City Biennale of Urbanism and Architecture (2012). She was awarded a National Endowment for the Arts grant for (Re)Stitch Tampa. Ensuing from this research platform was the publication *(Re)Stitch Tampa: Designing the Post-War Coastal American City through Ecologies*, published by ACTAR (2017). Bassett has lectured internationally, including in China, India, and South Korea. She is an invited professor every summer for the BIADW (Busan International Architecture Design Workshop) at Pusan National University in Busan, South Korea.

She is currently collaborating with the Delhi School of Planning and Architecture (SPA) on the project, "Reducing Risk, Raising Resilience: Recovering the Public Spaces of Shahjahanabad through Participatory Conservation and Ecological Urbanism." This design research is supported with an Insight Development grant from the Social Sciences and Humanities Research Council of Canada.

Petra Blaisse is the founder and lead designer of Inside Outside, an award-winning landscape architecture firm based in Amsterdam. The firm was founded in 1991 and deals with interior and exterior design including architecture, interior design, landscape architecture, exhibition, and textile design. In 1978, she went to work for the Stedelijk Museum in Amsterdam in the Department of Applied Arts. From 1986 onwards, Blaisse worked as a freelance exhibition designer. In realizing a series of award-winning installations and exhibitions with OMA (The Office for Metropolitan Architecture), Blaisse acquired an extensive knowledge on the use of light, composition, colour, materials, and acoustics to create environments that would impact her later work. Blaisse has lectured extensively in Europe, Asia, and the United States. Over the years, she has been a guest lecturer for the University of Tokyo, UCLA, Harvard University, Technical University in Delft, Cornell University, ARCAM, TEDxTilburg, and many other prestigious platforms and universities.

Jeffrey D. Blankenship is an Associate Professor of Architectural Studies at Hobart and William Smith Colleges in Geneva, NY, where he teaches the history of modern landscape architecture and design studios. His interests are in cultural landscape studies and design history within the broader context of twentieth-century intellectual history. He is currently working on a monograph for the LSU Press, *Everyday Modernity: J.B. Jackson and Landscape Magazine, 1951–1968*.

Ana Rita Sá Carneiro is a Graduate and Postgraduate Professor of Urban Development in the Department of Architecture and Urbanism, Federal University of Pernambuco (DAU/UFPE). She is a member of the International Committee of Cultural Landscapes ICOMOS-IFLA, representative of ICOMOS Brazil, coordinator of the Landscape Laboratory of DAU-UFPE (www.paisagem.net.br), member of ABAP, leader of the CNPq Research Group: Burle Marx Gardens, and researcher of the research groups: Landscape Thinking and Natural Heritage Conservation. Her research areas include landscape theory, historical garden, cultural landscape, conservation of cultural heritage, and public space.

Tina George is an MLA from the University of Guelph. She has a previous master's degree in Crop Sciences working extensively on genetic modification to field crops. During her study, she came to recognize that the most optimal future is in picking up the best pieces of the past. This knowledge of the power of the past and her passion for art styles influenced her research in Landscape Architecture, calling for the revival of collage techniques. Her inquiry of case studies suggests collage's unique ability in encouraging discourse and making it a more effective representational technique in praxis.

Adriaan Geuze is the founder and director of the three branches of West 8 in Rotterdam, North America, and Brussels and brings more than 35 years of experience in the fields of landscape architecture, urban design, and infrastructure. Geuze is internationally acclaimed for his pioneering work in the Netherlands and abroad. Under his leadership, West 8 has realized more than 200 projects in three decades. Many of them are the result of major international design competitions, including those for Governors Island in New York, Madrid Rio, Yongsan Park in Seoul, and the Waterfront Toronto in Canada. Geuze has received numerous awards for his work, including the Dutch Prix-de-Rome, the Maaskant Prize, the Oeuvre Prize for Architecture of the Mondriaan Fund, the ARC Oeuvre Award, and the ULI Netherlands Leadership Award. International prizes include the Veronica Rudge Green Prize for Urban Design (Harvard Design School) and several Honour Awards from the American Society of Landscape Architects. He frequently gives lectures and teaches at institutions worldwide.

Javier González-Campaña is a Partner and Principal at Balmori Associates. He joined the firm in 1998. Javier holds a Master of Science in Urban Ecology from Yale University's School of Forestry and Environmental Studies (2003) and a degree in Agricultural Engineering from the School of Forest and Agricultural Science at the Universidad Nacional de La Plata in Argentina.

Kona A. Gray, FASLA, PLA. As a Principal at EDSA with 27 years of experience in 30+ countries, his global design and management sense has positively shaped the outcomes of many environments. His portfolio includes large-scale planning and detailed site design with emphasis on communities, parks, hospitality, urban, and campus environments that solve meaningful issues. Currently, he serves as ASLA Representative to the Landscape Architecture Accreditation Board and recently served as ASLA Vice President Professional Practice. He is a Past President of the Landscape Architecture Foundation and active member of the Urban Land Institute. Kona earned a Bachelor of Landscape Architecture degree from the University of Georgia.

Alison Hirsch, PhD, FAAR, is Associate Professor and Director of both the Graduate Program in Landscape Architecture + Urbanism and the Landscape Justice Initiative at the University of Southern California. Alison's authored and edited books include *City Choreographer: Lawrence Halprin in Urban Renewal America* (University of Minnesota) and *The Landscape Imagination: Collected Essays of James Corner* (Princeton Architectural Press). She is co-founder and Research Director of Foreground Design Agency, a critical landscape practice. Alison is currently deeply immersed in research on the working landscapes of the San Joaquin Valley as a landscape of extreme inequality and risk, supported by the Landscape Architecture Foundation Fellowship for Innovation and Leadership (2020–2021).

Walter J. Hood is Professor and former Chair of Landscape Architecture at the University of California, Berkeley, and the Principal of Hood Design Studio in Oakland, California. He has educational background in landscape architecture and architecture (BLA 1981 at North Carolina A&T State University; MLA and MArch 1989 at UC Berkeley) and in fine arts (Distinguished MFA, 2013, at the School of the Art Institute of Chicago). His cultural practice engages landscape, urban and community design, art, and fabrication. Hood is a highly decorated professor and designer. He also has an extensive research portfolio and has numerous publications. In 2019, Walter Hood was awarded the Macarthur Fellowship. In 2010, he received the Cooper-Hewitt National Design Award for Landscape Design, and he was also a Fellow at the American Academy in Rome. More recently, he received the President's Medal for the Architectural League, 2021. Hood has influenced many students and landscape architects throughout the years. Hood has brought tremendous contributions to the landscape architectural profession.

Eelco Hooftman, together with Bridget Baines, is founder of GROSS.MAX. Landscape Architects. Between 1990 and 2008, he taught at Edinburgh College of Art and was co-founder of an innovative post-graduate program, Art, Space, and Nature. From 2008 to 2017, Hooftman was visiting studio instructor at the GSD, Harvard. He integrates theory and practice of landscape architecture in an extensive output in international projects and award-winning competition designs. The projects of GROSS.MAX. combine a Dutch sense of experimentation, a German sense of rigor, and a British sense of humour.

Noémie Lafaurie-Debany is a Partner and Principal at Balmori Associates. She joined the firm in 2007. She studied architecture at the École Nationale Supérieure D'Architecture De Nancy, France. Noemie was awarded a scholarship and spent her final year studying Nordic design and architecture at the University of Oulu in Finland.

David Malda is a Principal at GGN in Seattle. GGN was founded in 1999, by Jennifer Guthrie, Shannon Nichol, and Kathryn Gustafson. David leads planning and design projects throughout the United States, with an emphasis on the role of narrative and engagement to inform design. His focus on drawing and making throughout the design process is central to his role within the office and he shares this experience in numerous studio reviews and student engagements throughout the year. David holds a Masters in Architecture and a Masters of Landscape Architecture from the University of Virginia.

Matt Perotto holds an undergraduate degree in Planning with a specialization in Urban Design, and a Masters in Landscape Architecture from the University of Toronto Daniels Faculty where he graduated with Academic Honors and the Canadian Society of Landscape Architects Award of Merit. After his MLA, Matt joined Hargreaves Jones Landscape Architecture, where he has led a number of the office's landscape and master planning design projects, proposals, and international design competitions. As an Associate, he manages the production of design studies, text narratives, and presentation graphics, leading coordination processes with clients and sub-consultants and participates in public open houses and community engagement workshops. At the firm, Matt deals with BIM technologies using Revit and the integration of parametric computation tools directly within the studio's design process. Matt is also a Sessional Lecturer in Visual Communication at the Masters of Landscape Architecture program at the University of Toronto's John H. Daniels Faculty of Architecture, Landscape, and Design and is an Instructor and Lecturer in the School of Landscape Architecture and Design at the Academy of Art University in San Francisco.

Linda Pollak is a Principal of Marpillero Pollak Architects (MPA), a member of the NYC Design + Construction Excellence Program. MPA projects have received awards from the Architects Newspaper, the Center for Active Design, Architizer, the Chicago Athenaeum, American Institute of Architects, American Society of Landscape Architects, and others.

Linda is co-author, with Anita Berrizbeitia, of *Inside Outside: Between Architecture and Landscape*, and author of essays in books including *Reconstructing Urban Landscapes*, *Landscape Urbanism Reader*, *Large Parks*, and *Case: Downsview*, and journals. She serves on the Board of Directors of the Storefront for Art and Architecture and is a Contributing Editor for Lotus International. Her research on architecture and urban landscape has been recognized with grants and fellowships from the American Academy in Rome and many others. Linda is currently Adjunct Professor at The Cooper Union. She was a member of the Faculty of Architecture at Harvard Graduate School of Design from 1992 to 2004 and has taught at RISD, University of Pennsylvania, and Cornell.

Ken Smith is a professionally trained landscape architect with a background in design and fine arts. He holds professional degrees from Iowa State University and Harvard's Graduate School of Design. Ken Smith Workshop, founded in 1992, is based in New York City. His practice explores the relationship between art, contemporary culture, and the environment. He is committed to creating landscapes and public spaces that improve the quality of urban life. Smith has served as an adjunct professor, principally at City College of New York and Harvard's Graduate School of Design. He has been widely published in both the popular and academic press. He is a Fellow of the American Society of Landscape Architects. His work has received many awards including eight national ASLA awards, two Municipal Arts Society MASterworks Awards, and awards from the American Institute of Architects, American Planning Association, and the National Parks Service.

Frederick R. Steiner is dean and Paley Professor at the University of Pennsylvania Stuart Weitzman School of Design (Weitzman School) and co-executive director of its Ian L. McHarg Center for Urbanism and Ecology. Previously, he served for 15 years as dean of the School of Architecture and Henry M. Rockwell Chair in Architecture at The University of Texas at Austin. Before that, he taught at Penn and the following institutions: Arizona State University, Washington State University, and the University of Colorado at Denver. During 2013–2014, he was the William A. Bernoudy Architect in Residence at the American Academy in Rome. A fellow of both the American Society of Landscape Architects and the Council of Educators in Landscape Architecture and faculty fellow at the Penn Institute for Urban Research, he has written, edited, or co-edited 21 books. Dean Steiner earned a Master of Community Planning and a BS in Design from the University of Cincinnati, and his PhD and MA in city and regional planning and a Master of Regional Planning from the University of Pennsylvania. Dean Steiner also received an honorary M.Phil. in Human Ecology from the College of the Atlantic.

Halina Steiner is an Assistant Professor of landscape architecture in the Knowlton School at Ohio State University. Her research, Forensic Hydrology, focuses on overlaps between professional practice, hydrology, and infrastructure with an emphasis on scale and systems. This interest comes from a background in visual communication and her prior work as Design Director for DLANDstudio. Forensic Hydrology is a hybrid investigation including small-scale interventions to capture sediment from roads, temporary landscape installations, and narrative drawings challenging how we represent and understand water. Her research has appeared in *Landscape Architecture Magazine*, *LA Frontiers*, *JAE*, and the *Landscape Research Record*.

Chip Sullivan, FASLA, is an artist and Professor of Landscape Architecture and Environmental Planning at the College of Environmental Design, University of California, Berkeley. He has devoted his career to exploring and promoting landscape architecture as an intersection of art and ecology. Sullivan is the author of many popular titles in the profession, including *Drawing the Landscape*, *Garden & Climate*, and *Illustrated History of Landscape Design*, co-authored with Elizabeth Boults, ASLA. His latest book, *Cartooning the Landscape*, concerns the metaphysics of drawing and learning how to 'see.' The Foundation for Landscape Studies selected *Cartooning the Landscape* for the 2017 John Brinckerhoff Jackson Prize for accomplishment in the field of garden history and landscape studies. He is Fellow of the American Society of Landscape Architects and Fellow of the Council of Educators in Landscape Architecture.

Peter Walker has exerted a significant influence on the field of landscape architecture over a five-decade career. Educated at the University of California at Berkeley and the Harvard University Graduate School of Design, Walker has taught, lectured, written, and served as an advisor to numerous public agencies. Projects range from small gardens to new cities, from urban plazas to corporate headquarters and academic campuses. With a dedicated concern for urban and environmental issues, in a variety of geographic and cultural contexts, from the United States to Japan, China, Australia, and Europe. Walker is a Fellow of the American Society of Landscape Architects and the Institute for Urban Design and, in addition to numerous awards for specific projects, has been granted the Honor Award of the American Institute of Architects, Harvard's Centennial Medal, University of Virginia's Thomas Jefferson Medal, the ASLA Medal, and the IFLA Sir Geoffrey Jellicoe Gold Medal.

Jillian Walliss is an Associate Professor in landscape architecture at the University of Melbourne. Her research explores the relationship between theory, culture, and contemporary design practice. She has published widely, including the books *Landscape Architecture and Digital Technologies: Re-conceptualising Design and Making* (2016) and *The Big Asian Book of Landscape Architecture* (2020), both developed with Heike Rahmann.

Thaisa Way, FASLA, FAAR (BS UC Berkeley, MArchH UVa, PhD Cornell University) is the Program Director for Garden & Landscape Studies at Dumbarton Oaks Research Library and Collection, a Harvard University research institution located in Washington, DC. As an urban landscape historian, she teaches history, theory, and design at the College of Built Environments, University of Washington. Her publications foreground questions of history, gender, and shaping the landscape. Her book, *Unbounded Practices: Women, Landscape Architecture, and Early Twentieth Century Design* (UVa Press, 2009/2013), was awarded the J.B. Jackson Book Award. Her book *From Modern Space to Urban Ecological Design: The Landscape Architecture of Richard Haag* (UW Press 2015) explores the narrative of post-industrial cities and practices of landscape architecture. She co-edited with Ken Yocom, Ben Spencer, and Jeff Hou *Now Urbanism: The Future City Is Here* (Routledge 2014). *River Cities/City Rivers* (Harvard Press 2018) is a collection of essays contributing to urban and environmental history. Her book *GGN 1999–2018* (Timber Press 2018) is a foray into descriptions of design as a process. She is currently co-editing with Eric Avila a volume of essays on *Segregation and Resistance in the Urban Landscape*.

Acknowledgements

The Representing Landscapes book series has been over a ten-year journey of exploring the role and evolution of visual representation in landscape architecture. The prior books have dealt with the visualization works by students from various landscape architecture programs across the globe. This specific publication captures a reflection on the history of visual communication, mainly elevated by key influential landscape architects in the past 100 years. The book profiles approximately 25 historical to contemporary landscape architects, some with academic affiliations. Of course, there are many more landscape architects in recent history whom have had influential imprints in the profession pertaining to visualization. We were not able to profile more key figures at this point for various reasons. I am particularly grateful to have Professor Martin J. Holland in joining me on this journey. Martin specializes in landscape architectural history and offers a great compliment to the historical aspects of visual communication in the field. Thank you Martin for your valuable input and editing throughout the publication.

This book would not be possible without the personal and professional contributions from key landscape architectural scholars and the landscape architects profiled in this publication. I would like to acknowledge all the scholars and landscape architects who have provided valuable material towards this publication. Their contributions have revealed the personal and critical insights into the art of the visual messaging of their projects. I would like to express my gratitude to the following scholars and landscape architects: Halina Steiner and Frederick Steiner (it is lovely to have father and daughter in the same publication), Thaisa Way, Shannon Bassett, Ana Rita Sá Carneiro, Jeffrey D. Blankenship, Kona Gray, Chip Sullivan, Alison Hirsch, Javier González-Campaña, Noémie Lafaurie-Debany, Peter Walker, Linda Pollak, Petra Blaisse, Adriaan Geuze, Matt Perotto, Shannon Nichol, David Malda, James Corner, Tina George, Ken Smith, Damian Holmes, Jillian Walliss, Kirsten Bauer, Sacha Coles, Stephen Buckle, Walter Hood, Eelco Hooftman, and the staff from each firm who helped accelerate the process—Jack Petch, Merel Haenen, Amelia Starr, Regine Ong, Brooke Dexter, and many others.

I would also like to thank Sean Kelly, the Director of the School of Environmental Design and Rural Development at the University of Guelph, for his support towards landscape architectural research. I would also like to thank our former student and research assistant, Justin Luth, for his assistance in managing the contents of this book. We are grateful for his hard work during this process.

Thank you to Routledge (Taylor & Francis Group) for their ongoing support, marketing, and for the creative vision regarding this publication of the history of visual communication and the Representing Landscapes Series, in general. The Routledge team has been a strong support of this content and I thank them kindly for all their support throughout the year. Thank you to Kate Schell, who is the Senior Editor at Routledge, Taylor & Francis Group in the Planning, Landscape, and Urban Design department, and Routledge team for their assistance throughout this process and their assistance in shaping this idea. Finally, I am grateful to my parents for their ongoing support and encouragement, to Serena, Siena, Giuliano, Sofia, and Isabella; and to my husband, Haim, for his devotion and patience, which has made this process a positive experience.

Nadia Amoroso

I am honoured to be co-editor of this collected volume, *Representing Landscapes: One Hundred Years of Visual Communication*. My thanks to Kate Schell, a senior editor at Routledge in Planning, Landscape and Urban Design, for her unwavering support of this project and to the press for its ongoing commitment to the series. As an educator of landscape architecture, I have found the Representing Landscapes series a valuable resource to help inspire students with their design work. It is my sincere hope that this collected volume contributes to that legacy.

This work would not be possible without the various scholars, practitioners, educators, and designers contributing to the volume. Their dedicated efforts and desire to share their research and methods forms the content of this book. Nadia has identified each of you in her acknowledgements. I also thank you for your contributions completed on a tight timeline, made even more challenging because of the lasting implications of the COVID-19 pandemic.

I would also like to join Nadia in thanking Sean Kelly, the Director of the School of Environmental Design and Rural Development (SEDRD) at the University of Guelph, for his leadership and the support of research in landscape architecture. In addition, Justin Luth has been indispensable in the collection, organization, and maintenance of our materials. Justin, thank you for your excellent work.

Finally, I would like to express my gratitude to Nadia Amoroso for inviting me into this collaboration. Nadia has faithfully navigated the Representing Landscapes series to ensure that it reflects the latest developments within the field of landscape architecture. She could have selected any number of talented scholars and professionals as her co-editor, and I am immensely thankful for the opportunity to work with her.

Martin J. Holland

1 Introduction

Nadia Amoroso

Visual communication is a vital aspect of landscape architecture. Over the past century, influential landscape architects have helped guide and shape how we understand landscape architectural ideas through visual media. Drawings serve as research devices and presentational means. Illustrations are powerful tools that allow the landscape architect to tell their story, communicate their design, and persuade their audience.

This publication profiles a collection of key historical to contemporary landscape architects who have developed a specific graphical style, technique, or branding that has captured their audiences' attention throughout the years. The visual quality and overall 'look' of the drawings are compelling and evoke a sense of interest that captures the viewer's attention. How do we understand the drawings, perceive them, and connect with them? This publication examines the graphic styles and techniques over the past century, from prolific landscape architects that have helped shaped our profession.

The drawings crafted by many influential modern architects are graphically alluring and powerful in expressing their architectural visions. For example, the late-American Architect Lebbeus Woods' drawings on "radical deconstruction/reconstruction" evoke mystery, chaos, and imagination. The visuals seem to be inspired by Picasso's cubism and 1980s science fiction films. Similarly, the sketches of 'starchitect' Daniel Libeskind, specifically in his 'Micromegas' works, showcase strong linear geometry and a technical delicacy that reveals the hypothetical architectural spaces. These drawings are not only communication devices but also works of art. Peter Cook's colourful sci-fi, 'out of the world' illustrations of *Plug-In City* pull us into the mystery of these spaces. His drawings and collages could have readily been displayed in the Dadaism section of an art gallery.

Similar to the kinds of drawings the Cook and Libeskind have contributed to the modern era of architecture, this publication celebrates the drawings and visual communication devices of over 20 influential landscape architects over the past century. Do the drawings evoke a sense of emotion or imagination? Are the drawings worthy enough to be exhibited at internationally renowned art galleries or museums? The delicately crafted clay models produced by George Hargreaves and Kathryn Gustafson are works of art and, at the same time, communicate the proposed design. These models offer a sense of elegance and visual quality that draws the viewer. The beautifully sculpted landforms offer a visual quality that speaks to all audiences. The clay models offer a more tactile and immediate connection, though perhaps less actuate than computational form-making outputs.[1]

In the 1980s and 1990s, Walter Hood's 'cut and paste' colourful design proposals collaged and superimposed on a 'black and white' aerial base were quite inventive at the time. This technique offered a creative way of defining the existing landscape context (in black and white) juxtaposed with a design exploration through a vibrant coloured plan. Hood used black and white photos of the local community added to the drawing. He also tested airbrushing colour applications in his earlier drawings, found in his *Urban Diaries* collection. James Corner's 'map-drawings' of the later-1990s have redefined the way landscape architects think about maps. Mapping has become a creative process with a blending of art and technical measures. His map-drawings become instruments to understand

1.1
Collage, Ideogram, Greenport Harborfront, New York, 1996 by James Corner.

the site but also offer visual curiosity. Corner also emphasized the "eidetic" image in his expressive collage-style perspectives, capturing a vivid and abstract vision. His earlier collage-style perspectives employ abstract image pieces, a generalized depiction of place, rough extractions, and an ambiguous frame (elements coming out of the frame itself). This kind of visual expression allowed the viewer to 'think' critically about the designed space (Figure 1.1).

The late-French landscape architect Yves Brunier's crude colour-copied collage drawings presented a 'bipolar' tension between the medium and the design. The collages showcased a chaotic representation of the design while, at the same time, offered a sense of mystery. Thirty years later, the digital imagery crafted by Scottish-based landscape architecture firm GROSS.MAX. provides a sense of awe, mysteriousness, and uncanniness, which leaves the viewer with an uncertain vision of the landscape design. GROSS.MAX.'s digital collages, using Photoshop, exude a surrealistic aura. Their imagery is inspired by the works of Picasso, Ernst, Smithson, and Warhol. GROSS.MAX. was also intrigued by the collages of Archigram and Super Studio. These kinds of works reflect the image production of popular advertising art as a means to communicate messages. GROSS.MAX.'s photomontages are abstract, bizarre but elegant, with a high degree of imagination and indeterminacy. Figure 1.2, entitled "Glasshouse," by GROSS.MAX. depicts their typical visual communication style. According to GROSS.MAX., this image is about meditative tranquility and serenity. The composition is near-symmetrical; the vanishing point perspective is classical renaissance. The image captures a sense of anticipation.

This publication highlights the art, craft, and style of over 20 landscape architects throughout the past century. It celebrates the visual creativity of these landscape architects that have shaped our profession. Due to the time and scope of the publication and the amount of research required, only a limited number of landscape architects have been profiled. Understanding many accomplished landscape architects could have been discussed in this publication, please realize that some firms also declined to be profiled.

This publication includes the visualizations of the following landscape architects: Gertrude Jekyll, Beatrix Cadwalader Farrand, Fletcher Steele, Isamu Noguchi, Roberto Burle Marx, John

1.2
Digital photo montage, Glasshouse, 2020, by GROSS.MAX./Barbora Micovska.

Brinckerhoff (J.B.) Jackson, Edward D. Stone (EDSA), Thomas Church, Garrett Eckbo, Lawrence Halprin, Ian McHarg, Diana Balmori (Balmori Associates), Peter Walker, Yves Brunier, Petra Blaisse (Inside/Outside), Adriaan Geuze (West 8), George Hargreaves (Hargreaves Jones Landscape Architecture), Kathryn Gustafson (GGN), James Corner (Field Operations), Ken Smith (Ken Smith Workshop), Chip Sullivan (UC Berkeley), Walter Hood (Hood Design Studio/UC Berkeley), ASPECT Studios, and GROSS.MAX.

With a knowledge base of these selected historical landscape architects, landscape scholars and historians have contributed critical and narrative essays highlighting the value and purpose of these visuals in shaping the profession and how we "view" landscape design. Other chapter contributors include the actual founders or senior staff of some of those profiled. These personal accounts offer a raw and authentic overview of their visual communication style, brand, and techniques to convey their designs.

Note

1 Rieder, K. "Modeling, Physical and Virtual." In: M. Treib (Ed.). *Representing Landscape Architecture*. London, UK: Taylor and Francis, 2008, p. 176.

Reference

Rieder, K. "Modeling, Physical and Virtual." In: M. Treib (Ed.). *Representing Landscape Architecture*. London, UK: Taylor and Francis, 2008, p. 176.

2 A System of Expression
Writing and Making Landscapes of Gertrude Jekyll

Halina Steiner

Garden artist, painter, photographer, embroiderer, metalworker, horticulturist, and perhaps, most importantly, writer, Gertrude Jekyll had a collection of skills that allowed her to examine the world through many creative lenses.[1] Though best known as an artist-gardener (a term she coined for her own work), her legacy can be found in her writing.[2] Writing can be understood as a form of representation for Jekyll. With 14 books and over 1,000 articles published, she used writing as a tool to disseminate information on garden design to a larger audience. The primary audiences of Jekyll's writings were of a particular class of the Victorian population, those with enough space to implement her strategies as well as wealth to afford a gardener.[3] While her intention was not to write books for mass consumption, today, they are studied by a broad cross-section of landscape architects and garden enthusiasts. Although the scale of her designs was much larger than today's average gardens, the concepts embedded within them are transferable to smaller-scaled plots of land. Her writings bring the reader to her home in Surrey, England, allowing them to notice what she noticed; to be an observer of the small things, like subtle shifts in colour, texture, and time.

Collecting Skills

While Jekyll collected many skills throughout her life, painting was the first focus of her studies. The formal education she received, coupled with travels where she could paint (in oil and watercolour) and study landscapes, provided a foundation for her later work with plants. Jekyll's experimentation with different painting media and her interest in the emerging field of colour theory greatly influenced her later approach to garden design and documentation.

Painting requires the artist to envision the final result before the brush makes its first stroke. In watercolour, lighter washes of colour are applied to the paper. When applying a wash of colour, perhaps blue, one must know what colours will be applied next; for example, a second wash of red will provide a purple colour where the washes overlap. One must understand the choreography of the painting in advance of applying the colour.

In oil painting, there are many techniques. Impasto, used by J.M.W. Turner, whose paintings Jekyll studied, applies heavy layers of paint, applied by brush or palette knife. The result is a painting rich with textures that catch the light at different angles created by the layers of paint. Turner's work demonstrated ideas on colour theory that appear in Jekyll's planting. However, much of Jekyll's approach to colour theory can be demonstrated by Michel-Eugène Chevreul's Law of Simultaneous Contrast,[4] which states:

> I beg the reader never forget when it is asserted of the phenomenon of simultaneous contrast, that one colour placed besides another receives such modification from it… that the two colours… have a mutual action.[5] (Figure 2.1)

2.1
Chevreul's colour wheel from *The Principles of Harmony and Contrast of Colours*.

Chevreul, a French chemist, made this discovery while working at the dying laboratory at Manufacture Royal des Gobelins, where complaints of dull fabrics were attributed to the dying process.[6] However, the root cause was that they were woven together with hues in close proximity, resulting in a dull appearance. If instead placed by contrasting colours, the result is bright, vibrant fabrics.[7] He published *The Principles of Harmony and Contrast of Colours* in 1839, and by 1854, it had been translated into German and English.[8] Chevreul further studied the application of contrasting colours in painting, tapestry, carpet, mosaic, stained glass, textile fabrics, typography, *enluminures*, architecture, dress, and horticulture.[9] Chevreul's theories extended beyond the silos of science. Before his book was even published, through many years of public lectures, his work quickly influenced tradesmen and artists alike. The public knowledge of his theory was first adapted by artisans and people working in applied arts and later by artists, including the neo-impressionists.[10] Chevreul's colour theories are evident in the work of painters from other movements such as Eugène Delacroix, Camille Pissarro, Claude Monet, and Paul Cézanne, all of whom Jekyll admired.[11]

Though one can read an impressive volume of books and articles written by Jekyll and subsequent books about her, it can still be difficult to uncover the roots of her design thinking.[12] It is important to remember the breadth of creative work she pursued and other techniques that may have also played a role. Her personal relationships also certainly influenced her approach. Her willingness to cultivate a diverse set of artistic skills was akin to her friend William Morris, who similarly thought about colour theory in both his craft and his poetry.[13] Furthermore, her association with gardener and writer William Robinson and frequent collaboration with architect Edwin Lutyens also played important roles in her garden designs.

While it seems plausible that all of her creative endeavours influenced her garden designs to some degree, central to these undertakings was painting. The techniques from painting are readily visible in Jekyll's approach to garden design. She understands the picture as a whole before she begins. Understanding the seasonal layers then applying them to the garden is akin to watercolour. Aspects, such as the colour, the details, the light, and the texture of each plant in the garden, are similar to

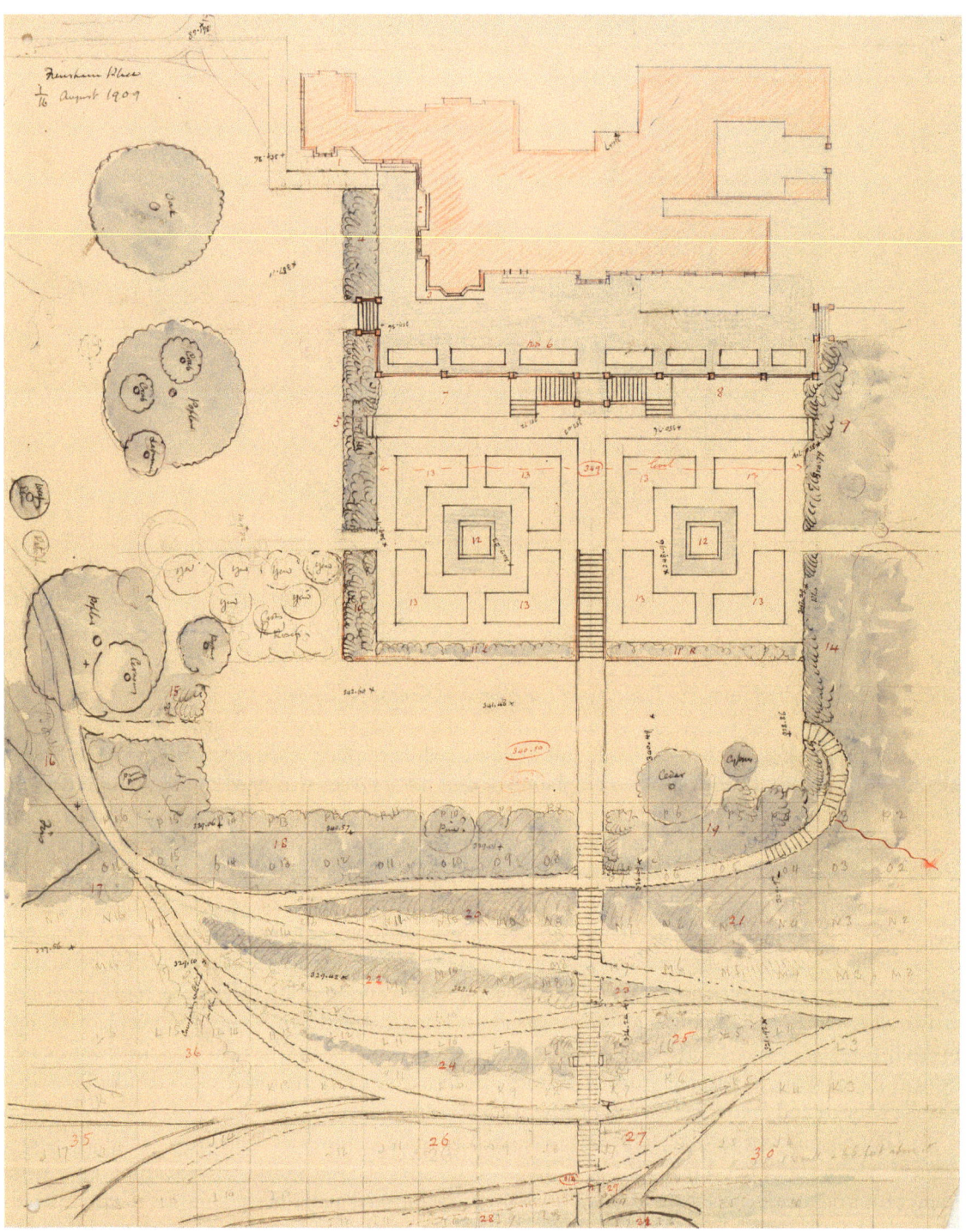

2.2
Plan of Frensham Place, 1909.

painting with oils. How do the colours of the blooms, the shape, and texture of the leaves work with the other plants and interact with light?

> She sought to draw the best out of every plant, using them and soil as paints and paper. She did not grow plants just for the sake of growing them, but only when they would add to the picture she had in her mind.[14]

When her diminishing eyesight required her to abstain from tedious work such as embroidery and painting, writing and garden designs became new outlets for creative expression.[15] She supplemented these endeavours with photography. She took photos of plants and planting beds for her own study and for use in her publications. In 28 years, she filled volumes of photo notebooks with over 2,100 images.[16] Michael Van Valkenburgh writes that Jekyll's photographs

> are seen as a collection of design speculations—recorded by camera rather than pen and sparsely labelled with spontaneous margin notes—we find that they are a part of her meticulous studies that join horticultural appropriateness with the visual and spatial concerns of landscape design.[17]

Jekyll chose to represent the landscape with words, photographic study, and the laboratory of her garden. Her plans for complex gardens are often simple pen and ink drawings. Hints of watercolour appear periodically. The plan at the Frensham Place hints back to her earlier interests; washes of watercolour paint are used to highlight plant massings that define spaces (Figure 2.2). Occasionally, reds and oranges appear on a planting plan, as shown at West Dean Park (Figure 2.3). The use of the pen to delineate planting drifts replaces the paintbrush at Brambletye (Figure 2.4). Once scientific names are added to the drifts, the texture and intricacy of the planting bed contrast with the openness of the lawn. In the context of these relatively simple drawings, Jekyll's writing assists us in understanding the rationale for planting combinations and strategies.

2.3
Herbaceous border plan for West Dean Park, 1898.

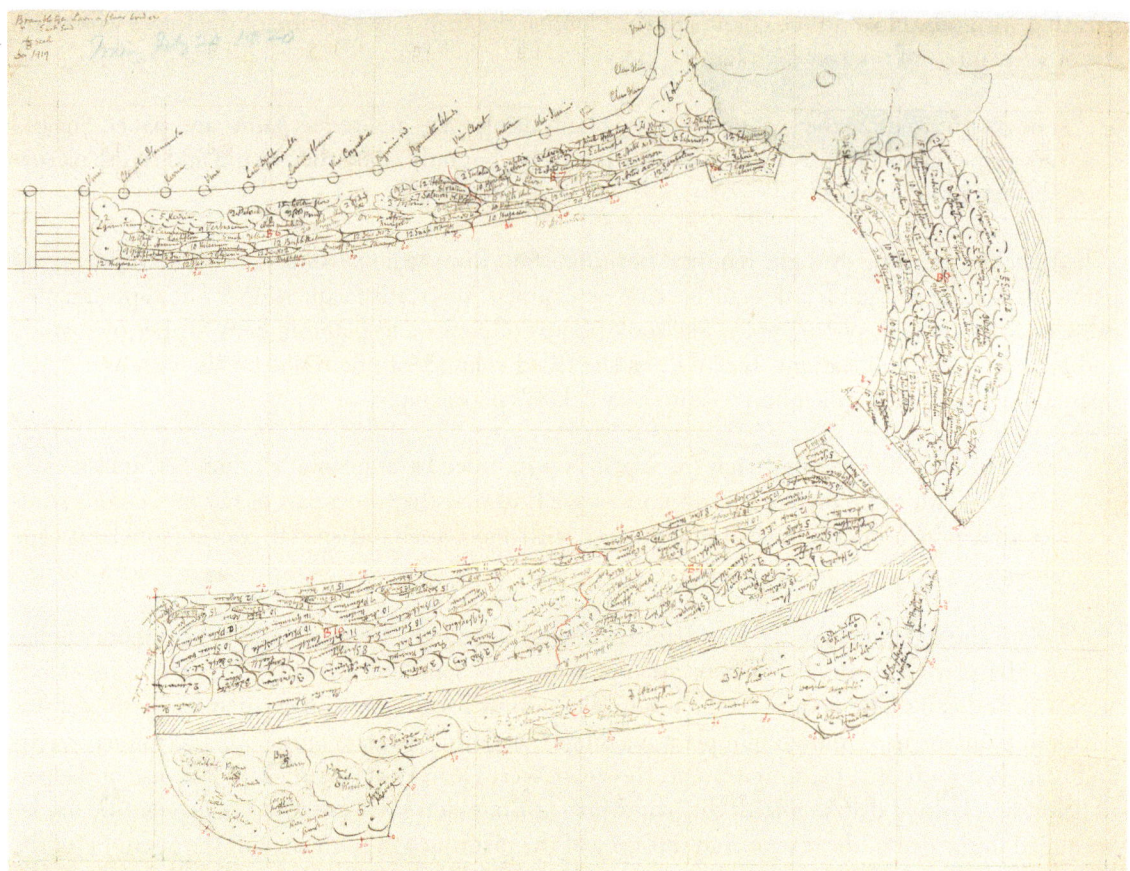

2.4
Lawn and herbaceous border at Brambletye, 1919.

Writing

One not familiar with Jekyll's writing may assume that her books are filled with facts, drawings, and photographs of plants—and, to a certain extent, they are. However, plants are not arranged alphabetically, accompanied by their mugshot, growth statistics, and their common and scientific names. Rather, Jekyll unveils the plants as an experience. In reading her work, it is almost as if you are at a party and being introduced to each plant as you move through the garden. For example, Jekyll may say, let me introduce you to the Garland Rose, who "every summer is loaded with graceful wreaths of bloom."[18] To illustrate how the rose is maintained, she explains:

> it has never been trained or staked, but grows as a natural fountain; the branches are neither pruned nor shortened. The only attention it receives is that every three or four years the internal mass of old dead wood is cut right out, when the bush seems to spring to new life.[19]

She affords this same generosity to annuals in *Annuals & Amps: Biennials, the Best Annual and Biennial Plants and Their Uses in the Garden*. When she writes of hedge-forming and climbing annuals:

> Of these the first that come to mind are the Sweet Peas, and for a whole continuous hedge nothing can be more delightful. The range of colouring is now so extensive, and the variety of form and marking so distinct, that one may walk along a considerable length of hedge without coming to an end of the pleasure and interest. There is no other climbing plant of which so much may be said, though some others may be satisfactorily used as hedges.[20]

Even when a catalogue of plants is called for, the personality of the plant still shines through: "Linaria maroccana, white variety; h.a.; 12 inches.—A charming plant. It delights in a warm, sunny spot, and

A System of Expression 9

is useful, grown in pots, for the conservatory."[21] Narrative descriptions of the spatial characteristics as one moves through the gardens are coupled with seasonal observations that thread together the detailed plant descriptions.

The writing provides the same honest account of all plants she introduces. This gives the reader a richer understanding of how a plant may be used in a new composition. As if to say, white colour provides such a good base for blue, but it also can be great if used in this way. There is no single way to use a plant, but one must first understand its obvious and hidden potential. Such understanding enables us to celebrate the emergence of plants, their showiest of moments, and the beauty in their decline and the emergence of something new in their place. Her approach to writing is painterly; she uses words to make a sketch of a garden bed. These are further supplemented with photographs and planting plans.

In the August 26, 1882 issue of *The Garden*, in an article titled "Colour in the Flower Garden," she writes:

> One of the most important points in the arrangement of a garden is the placing of the flowers with regard to their colour-effect, and it is one that has been greatly neglected. Too often a garden is an assemblage of plants placed together haphazard, or if any intention is perceptible, as is commonly the case in the bedding system, the object aimed at is as great a number as possible of the most violent contrasts, with the result of a hard, garish vulgarity. Then, in the case of mixed borders, what is usually seen is either lines or evenly distributed spots of colour, wearing and annoying to the eye, in no way interesting, and proving only how poor an effect can be got by the misuse of possibly the best materials. Should it not rather be remembered that in setting a garden we are painting a picture, only it is a picture of hundreds of feet or yards instead of so many inches, painted with living flowers and seen by open daylight, so that to paint it right is a debt we owe the beauty of the flowers and the light of the sun. Therefore the colours should be placed with careful forethought and deliberation, and not dropped done in lifeless dabs, as he has them on palette.[22]

The Bank of Early Bulbs helps to clarify hair planting strategy (Figure 2.5). With roughly 25 plant species, one may not think of the strategy as simple. However, the structure of the design consists of two key

2.5

The Bank of Early Bulbs. From top to bottom, plan in March, April, May, and June. The bank of early bulbs features: white hyacinth, *muscari, anemone blanda, erythronium, fumaria, megasea, trientalis, scilla amoena, scilla siberica, chionodoxa, scilla bifolia, asnite crocus, puschkinia*, white crocus, *valerian, narcissus princeps, narcissus minor, narcissus nanus*, daylily, *narcissus pallidus praecox*, red hellebore, and ferns (Jekyll, G. "A March Study and the Border of Early Bulbs." *Op. cit.*).

components: anchor plants and early spring plants. Spring plants, composed of bulbs, give early springtime a burst of colour. Because the bulbs do not bloom at the same time, their placement and bloom time are carefully choreographed to provide the intended colour "the right effect in each grouping."[23] Bulbs are grouped by colour; within each grouping, subtle shifts in height and texture add to the effect. Drifts of ferns and hellebore anchor the space throughout the summer, fall, and winter. Jekyll writes:

> Through April and May the leaves of the bulbs are growing tall, and their seed-pods are carefully removed to prevent exhaustion. By the end of May the Ferns are just throwing up their leafy crooks; by June the feathery fronds are displayed in all their tender freshness; they spread over the whole bank, and we forget that there were any bulbs between. By the time the June garden, whose western boundary it forms, has come into fullest bloom it has become a completely furnished bank of ferns.[24]

Imagine this strategy with 75 flowering plants blooming throughout the summer months, forming a rich herbaceous border as is the case at Munstead, her home in Surrey. The 200-foot planting bed reveals the complexity of these design ambitions.[25] Jekyll assumes the role of a conductor—knowing when each flower would bloom, when it needed to be trimmed, what could fill in any holes: all evidence of an intimate knowledge of each plant. The resulting effect includes layers of plants to form a composition appearing as a wall of flowers, shifting drifts of colours as one moves along the path. The base of plants, unless used for striking foliage, is hidden by neighbouring plants. Plants closest to the path are lowest; as the plants move further away from the path, the height increases slightly with each new drift of plant until reaching 5–6 feet (Figure 2.6).

Jekyll's writing reveals a focus on time and experimentation. She provides readers with agency to approach their own garden and grants them permission to fail. She does this by acknowledging her own process in garden design: "It was a puzzle for many years to know how to treat these early bulbs, but at last a plan was devised that seems so satisfactory that I have no hesitation in advising it for general adoption."[26] Sally Festing's biography of Jekyll simply states:

> The power of Gertrude's books is in alerting readers to 'quicken the inventive faculty,' use no tight rules, but look, deliberate and then experiment. Gardening's perpetual challenge was one of its chief attractions. Each scrap of ground or water surface, every six inches of soil, provided opportunity.[27]

Gertrude Jekyll's writings were not dry accounts of her designs. She afforded the same amount of creative expression in her prose as she did the gardens themselves. With an active and approachable voice, she paints vivid pictures of her gardens, with the plants as dynamic subjects. It is perhaps this very 'approachability' that has made her works so enduring. A *New York Times* article from 2000, "IN THE GARDEN: Discovering Borders that Blend, a la Monet," states, "Gertrude Jekyll may be the Elvis of the gardening world. Gardeners from Trenton to Tokyo speak her name in reverent tones, and make pilgrimages to Munstead, her English home." While living, she had a wide influence on garden

2.6
Planting plan and sections through herbaceous border at Munstead show the wall of flowers effect.

design through the weekly publication *The Garden*, her books, and other journal articles. Her work has even been cited as an inspiration for Monet's own gardens.[28] Today, we can see evidence of her influence on the New Perennial movement, where drifts of plants continue to create seasonal splendour. While one could debate if Elvis is Jekyll's music equivalent, it is evident that her legacy is lasting.

Notes

1. Van Valkenburgh, M. "The Flower Gardens of Gertrude Jekyll and Their Twentieth-Century Transformations." *Design Quarterly*, 1987, 137: 8.
2. Bisgrove, R. "The Colour of Creation: Gertrude Jekyll and the Art of Flowers." *Journal of Experimental Botany*, 2013, 64(18): 5785.
3. Festing, S. *Gertrude Jekyll*. London, UK: Viking, 1991, pp. 169–170.
4. Brown, J. *Gardens of a Golden Afternoon*. London, UK: Penguin Books Ltd, 1982, pp. 41–43.
5. Viénot, F. "Michel-Eugène Chevreal: From Laws and Principles to the Production of Colour Plates." *Color Research & Application*, 2002, 27(1): 5.
6. Ibid., pp. 4–5.
7. Ibid., p. 5.
8. Roque, G. *Chevreul's Color Theory and Its Consequences for Artists*. London, UK: The Colour Group (GB), 2011, pp. 11–21. Web. 15 May 2021. https://www.colour.org.uk/wp-content/uploads/2017/10/Chevreuls-Law-F1-web-good.pdf.
9. Viénot, *op. cit.*, p. 5.
10. Roque, *op. cit.*
11. Brown, *op. cit.*, p. 42.
12. Van Valkenburgh, *op. cit.*, p. 3.
13. Byron, K. "A Study of the Color and Imagery of the Poetry of William Morris" 1957, (pp. 1–117). Web. 10 May 2021. https://scholarship.richmond.edu/cgi/viewcontent.cgi?article=1121&context=masters-theses.
14. Thomas, G. "The Influence of Gertrude Jekyll on the Use of Roses in Gardens and Garden Design." *Garden History*, 1977, 5(1): 57.
15. Batey, M. "Landscape with Flowers: West Surrey: The Background to Gertrude Jekyll's Art." *Garden History*, Spring 1974, 2(2): 15.
16. Van Valkenburgh, *op. cit.*, p. 4.
17. Ibid., p. 8.
18. Jekyll, G. "The June Garden." In: The Estate of Gertrude Jekyll (Ed.). *Colour Schemes for the Flower Garden*. Woodbridge, UK: Baron Publishing, 1982, pp. 101–123.
19. Ibid.
20. Jekyll, G. *Annual & Biennials, the Best Annual and Biennial Plants and Their Uses in the Garden*. London, UK: Country Life Ltd., 1916, p. 45.
21. Ibid., p. 116.
22. Jekyll, G. "Colour in the Flower Garden." *The Garden*, August 5, 1882, XXII: 177.
23. Jekyll, G. "A March Study and the Border of Early Bulbs." In: The Estate of Gertrude Jekyll (Ed.). *Colour Schemes for the Flower Garden*. Woodbridge, UK: Baron Publishing, 1982, pp. 21–35.
24. Ibid.
25. Brown, *op. cit.*, pp. 41–43.
26. Jekyll, G. "A March Study and the Border of Early Bulbs." *op. cit.*
27. Festing, *op. cit.*, p. 157.
28. "The Art of Monet's Garden." *Country Life*, January 15, 2016. Web. 5 May 2021. https://www.countrylife.co.uk/luxury/art-and-antiques/the-art-of-monets-garden-82327.

References

"The Art of Monet's Garden." *Country Life*, January 15, 2016. Web. 5 May 2021. https://www.countrylife.co.uk/luxury/art-and-antiques/the-art-of-monets-garden-82327.

Batey, M. "Landscape with Flowers: West Surrey: The Background to Gertrude Jekyll's Art." *Garden History*, 2(2) (1974): 15.

Bisgrove, R. "The Colour of Creation: Gertrude Jekyll and the Art of Flowers." *Journal of Experimental Botany*, 64(18) (2013): 5785.

Brown, J. *Gardens of a Golden Afternoon*. London, UK: Penguin Books Ltd, 1982, pp. 41–43.
Byron, K. "A Study of the Color and Imagery of the Poetry of William Morris" 1957. (pp. 1–117). Web. 10 May 2021. https://scholarship.richmond.edu/cgi/viewcontent.cgi?article=1121&context=masters-theses.
Festing, S. *Gertrude Jekyll*. London, UK: Viking, 1991, pp. 169–170.
Jekyll, G. "A March Study and the Border of Early Bulbs." *Colour Schemes for the Flower Garden*. London, UK: Country Life Ltd., 1982, pp. 21–35.
Jekyll, G. "Colour in the Flower Garden." *The Garden*, August 5, 1882, Vol. XXII: 177.
Jekyll, G. "The June Garden." In: The Estate of Gertrude Jekyll (Ed.). *Colour Schemes for the Flower Garden*. Woodbridge, UK: Baron Publishing; , 1982, pp. 101–123.
Jekyll, G. *Annual: Biennials, the Best Annual and Biennial Plants and Their Uses in the Garden*. London, UK: Country Life Ltd., 1916, p. 45.
Roque, G. *Chevreul's Color Theory and Its Consequences for Artists*. London, UK: The Colour Group (GB), 2011, pp. 11–21. Web. 15 May 2021. https://www.colour.org.uk/wp-content/uploads/2017/10/Chevreuls-Law-F1-web-good.pdf.
Thomas, G. "The Influence of Gertrude Jekyll on the Use of Roses in Gardens and Garden Design." *Garden History*, 5(1) (1977): 57.
Van Valkenburgh, M. "The Flower Gardens of Gertrude Jekyll and Their Twentieth-Century Transformations." *Design Quarterly*, 137 (1987): 8.
Viénot, F. "Michel-Eugène Chevreal: From Laws and Principles to the Production of Colour Plates." *Color Research & Application*, 27(1) (2002): 5.

3 Beatrix Farrand
Representing Landscape in Prose and Drawings

Thaisa Way

Born in 1872, Beatrix Cadwalader Jones (later Farrand) grew up in an era when opportunities for women were expanding for access to colleges and to careers including those in art, botany, and teaching, among others. Landscape architecture was one of the fields attracting women, likely due to its emergence as a profession at the same time as women were entering the workforce, but more importantly, likely, because it fit well with the education and training offered to women of the era.

As with her Victorian-era upper-class female peers, Farrand was tutored in French, German, Italian, and Greek as well as music and art. While music was an option, she met Charles Sprague Sargent, curator of Harvard's Arnold Arboretum, in 1892, who suggested that she study landscape gardening. Farrand made several long-term visits between 1893 and 1895 with the Sargents at their home, Holm Lea, in Boston. Farrand published her first article in Sargent's journal, *Garden and Forest*, in the fall of 1893 as a letter to the editor about landscape gardening with native trees of Bar Harbor, Maine.[1]

In 1895, Farrand, with her mother as a companion, travelled to Europe to study the great gardens and to "learn from all the great arts as all art is akin."[2] She kept extensive notes, primarily descriptions and critiques of what she saw in the gardens of France, Italy, and England. On her return to New York City in 1896, Farrand launched her professional practice with an office located on the top floor of her mother's house on East 11th Street. Her first commissions included a project for William Garrison in Tuxedo, New York, and W.H. Bliss, a New York banker, and his wife, Anna Blakesley Bliss, tackling the drainage on their property in Bar Harbor, Maine. In January of 1899, three years after opening her office, Farrand was invited to join Samuel Parsons and other invited practitioners to form the American Society of Landscape Architects (ASLA). This acknowledgement of her practice—although she was still relatively new to the field—would be important as she sought to expand her practice.

Over the course of five decades, Farrand led an active practice with commissions across the country, only limited by the number she could personally oversee. The repertoire included botanical gardens, university campus plans, and public gardens. She was responsible for the design and implementation of campus projects for Princeton University (1912–1943); Yale University (1922–1945); Hamilton College (1924); Vassar College (1926–1927); California Institute of Technology (1928–1938); The University of Chicago (1929–1936); Pennsylvania School of Horticulture for Women (1931–1932); Occidental College (1937–1940); and Oberlin College (1939–1946) as well as the Rose Garden for the New York Botanical Garden in the Bronx Park (1915–1916) and three important garden designs: Dumbarton Oaks, The Eyrie, and the Willard Straight garden. She died in 1959 at Garland Farm, where she had moved after Reef Point to live her last years.

Dumbarton Oaks came to Farrand at the height of her career, when Mildred and Robert Bliss commissioned her to design the landscape and gardens for their new home in Washington, D.C., on the outskirts of Georgetown. The archives record over 1,226 drawings and sketches as well as a rich correspondence developed for the landscape and its garden rooms from 1922 to 1947. Drawing from the extensive archives, it is clear that she used both prose and drawings in her communication of design to her clients and patrons. Furthermore, she relied on her capacity to oversee in person the

DOI: 10.4324/9781003183402-3

actual construction and planting of her designs so that she might alter and refine the designs in place. In a manner of speaking, Farrand's toolbox of approaches included her facility with prose, whether in a long descriptive letter or a to-the-point telegram; drawings, including sketches and vignettes as well as detailed plans, engineering drawings, and full-scale delineations; and her oversight on site during construction and over decades of stewardship. These three approaches—prose, drawing, on-site refinements—to representation and communication of the design process informed Farrand's relationship with both her clients and the gardeners. The approaches also shaped the designs in their development as complex shapings of garden rooms in expansive landscapes, envisioned as works of living art.

Farrand understood the importance of a well-executed drawing to communicate with a client. Robert Patterson noted that she regretted that she never learned to draw[3]; it is more likely, however, that she could not draw as well as she would have liked. However, there are multiple examples of drawings early in her career, when it is more than likely she drew them: many in pencil on trace—an approach she would sustain throughout her career—as well as a number of watercolours. Like many contemporary landscape architects, including Frederick Law Olmsted, Farrand may have relied on employees to aid in the translation of her vision into visual presentations. As a practice, drawing was

3.1
Garden for Witherell (Nathaniel), Greenwich, CT, 1899–1900, watercolour on heavy paper. This is an early project revealing Farrand's use of watercolour as a good media for communicating the character of the garden. Image from Beatrix Jones Farrand Collection, 1866–1959, UC Berkeley, Environmental Design Archives.

central to her design process, and her firm created thick recordings of the design in drawings, whether by her hand or a staff designer, including sketches, sections, plans, vignettes, details, and, for selected projects, watercolours. Presentations were usually pencil on trace. Full-size detail drawings were produced on trace paper with notes as needed to communicate the materials and specifications. Study drawings were beautifully detailed without suggesting that they were the works of art. Watercolour paintings of selected gardens were done on heavy paper, although not framed or otherwise presented as a work of art (Figure 3.1). Farrand clearly understood the drawings to be a way to communicate her ideas for the physical landscape and garden design.

Farrand developed every design in collaboration with her clients, particularly Mildred Bliss. She shared her iterative process, often begun in prose and then in sketches. Farrand would explain the design question and its context, followed by a description of possible responses. In a report on Dumbarton Oaks, Farrand suggested that "the north slopes of the property should properly be studied from the ground itself rather than from any plan, as the contours and expressions of the ground will control the plantations more strongly than any other feature."[4] She understood the need to combine site visits as well as inventories of the site and professional surveys. Once these were complete, she would offer a vision for the landscape in prose. The textual vision was then augmented with a series of sketches and drawings for which she would present the benefits and challenges of each decision, only sometimes noting "designer's preference." She would then mark the drawing based on the reaction by Bliss. As Bliss described:

> Such were Mrs. Farrand's integrity and loyalty that…never in all the years did she impose a detail of which she was 'sure' but which the owners did not 'see' and never were the owners so persuasive as to insist on a design which Mrs. Farrand's inner eye could not accept.[5]

A beautiful example of the power of Farrand's prose is her letter of June 24–25, 1922, in which she provided Bliss with a description of her vision for the Dumbarton Oaks landscape. Bliss responded to Farrand with the note, "Your letter and its enclosures have made us purr with contentment…You have got it exactly; in every respect, and I can't be patient until you get back here and start to realize your and our mutual dream."[6] In this letter, which included no sketches, Farrand laid out a vision that drew from her walks in the landscape and discussions with Bliss, and this vision would frame the construction of the landscape over the next 25 years. The landscape remains a manifestation of that early vision laid out so clearly in prose, as it guided design decisions as well as the stewardship of the work over the next century.

Drawings were nonetheless important to Farrand as she developed and refined each design. Farrand began each project by walking with her clients through the landscape and then having it surveyed. Her attention to topography is evident in her first drawings for the landscape, often overlaid on the contour lines (Figure 3.2). In addition to these drawings, Farrand produced engineering drawings of the Dumbarton Oaks landscape that detailed the cut and fill necessary to build the garden terraces along the steep slope. She carefully calculated the height and breadth of the retaining walls and the dimensions of the stairs to create the flat terrace that would feature the Rose Garden and to create the experience of a relaxed stroll down what was, in fact, a steep slope.

Farrand also used drawings to offer alternatives. In her 1922 report, Farrand offered Bliss four Rose Garden proposals, "all modifications of the same design in three parts."[7] The first, Farrand noted, was "developed symmetrically in the familiar Italian and Eighteenth-Century English design of the rectangle with the circle in the centre." The second design was not symmetrical, while the third omitted the rose beds of the first plan, "throwing this space into the same Italianate design, which of course has been used since the days of Babylon." The fourth plan was the simplest in Farrand's opinion. She asked Bliss to consider all of these alternatives, noting that "the whole thing is fluid and I want your help in arranging it"[8] (Figure 3.3). In a similar manner, Farrand offered Mrs. Abby Aldrich Rockefeller options for the gate for her Eyrie Garden on Seal Harbor in Maine (Figure 3.4). In each of these examples, the drawings are on trace in pencil, a media that Farrand used often, suggesting that while the details were well articulated, it remained a draft, an idea to be considered. There is a care to the details of proportion, scale, and relationship of the parts that are evident in the drawings

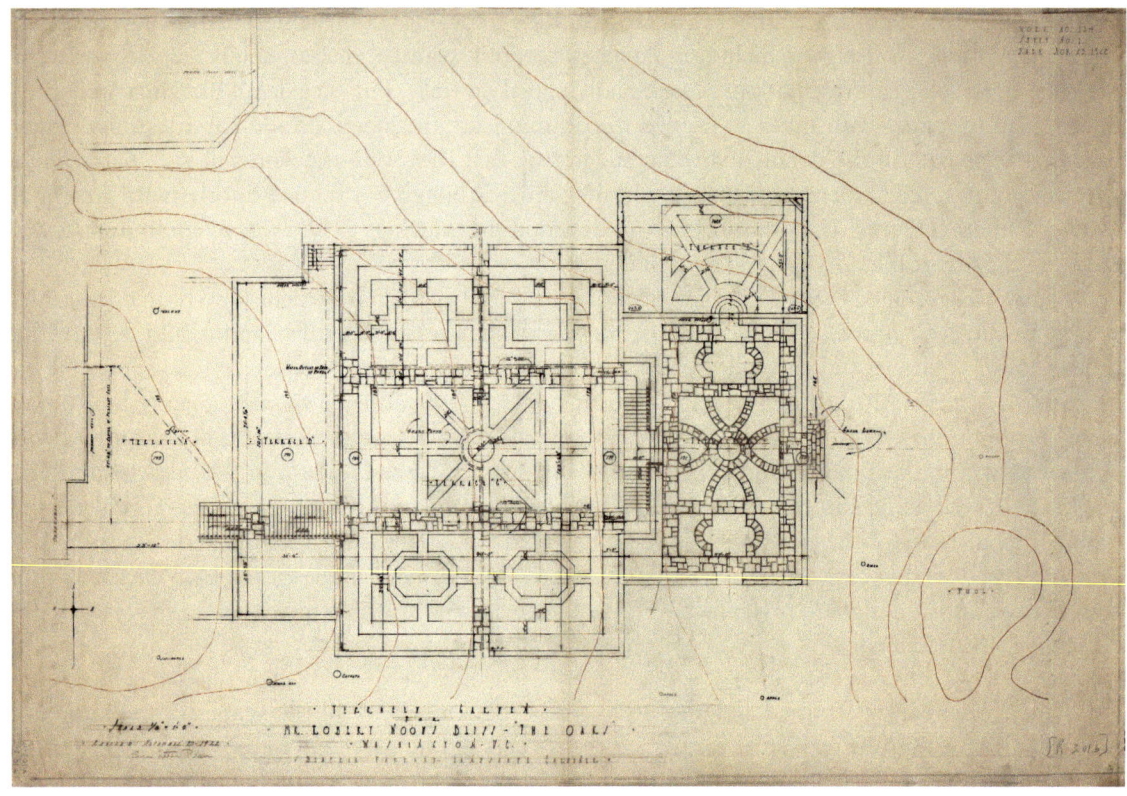

3.2
East garden terraces drawn over contour lines revealing the drop of almost 50 feet from the Orangery to Lovers Lane Pool. In this manner, Farrand directed her client's attention to the reasons for the retaining walls, the extent of the terraces, and the position of stairways. Image from Dumbarton Oaks Garden Archives, Dumbarton Oaks Research Library and Collection.

that Farrand does not reference in the prose, suggesting that she viewed the drawings as more than illustrations of the text, offering their own visual argument.

The design of the Arbor Terrace was one of the greatest challenges for Farrand, as she envisioned an intimate garden room overlooking the landscape below, from the orchard, to the herbaceous border, to the kitchen gardens and then the ravine and its woodland beyond. This garden room, like the Rose Garden, would require a large retaining wall on the western edge. Farrand determined to place a French inspired arbour along the western edge of the terrace both obscuring the massive stone wall and offering a Romantic Garden room within the garden room. For this project, she provided several vignettes (Figure 3.5). Another drawing on trace with pencil, Farrand again is making a suggestion for the character of the design and offering the details that might enhance the concept.

Drawings were pertinent to Farrand's interrogation of the human scale as it would be experienced in the garden, in relationship to the structural armature necessary to create the garden rooms. In a section drawing of the Rose Garden Wall (Figure 3.6), Farrand describes the scale of the retaining wall to explain why she is inserting a middle terrace into the landscape between the Beech terrace and Rose Garden. The use of the section allows Farrand to highlight the relationships of the parts to one another in scale and position. It is a visual language that might communicate the important design elements more clearly than any prose might. In other drawings, Farrand understood the power of colour, using it carefully. In a watercolour vignette for the stairs to the swimming pool, Farrand applies colours that differentiate between the stonework and the plants (Figure 3.7). Her attention to media and colour reflected the ideas Farrand sought to communicate, whether it was about scale, texture, seasonal colour, or differentiating architecture from plants.

Farrand's office also produced multiple full-scale drawings of details, especially as concerned the ornamentation (Figure 3.8). These drawings could be viewed in the office, on a table, or at the site held

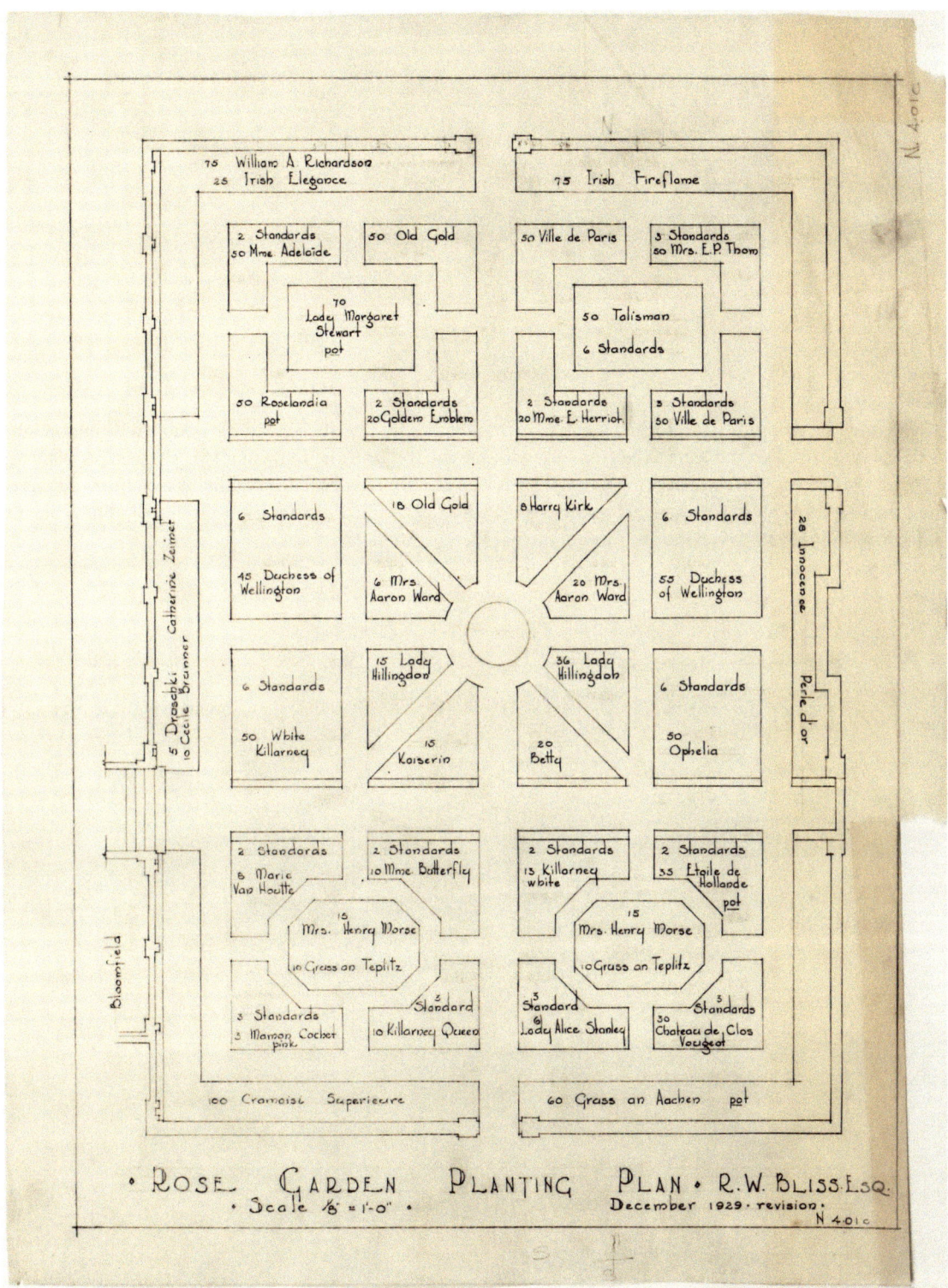

3.3
Rose Garden plan #4.01c. This is one of four alternative plans Farrand offered to Bliss for the layout of the Rose Garden. Each of the plans offered a slightly different colour layout as well as modified rose varieties. Image from Dumbarton Oaks Garden Archives, Dumbarton Oaks Rescarch Library and Collection.

3.4
Eyrie Garden for Abby Aldrich Rockefeller, a series of sketches for alternatives for the double wall, moon, and bottle gates, 1926. Farrand shows the first idea without the tree and the following two alternatives with the trees that are in place. On trace paper with graphite as well as coloured pencil for highlights in yellow and green, strengthening the ways in which plants would engage with the architecture. Image from Beatrix Jones Farrand Collection, 1866–1959, UC Berkeley, Environmental Design Archives.

3.5
Sketch of the pool and River God sculpture under the Arbor on Terrace E. For this drawing, Farrand depicted the plan in the upper right corner revealing the symmetry and simple nature of the design, while the vignette is drawn in a romantic style with its wooden arbour structure enclosing a flowing fountain and pool surrounded by cattails and other water-associated potted plants. Image from Beatrix Jones Farrand Collection, 1866–1959, UC Berkeley, Environmental Design Archives.

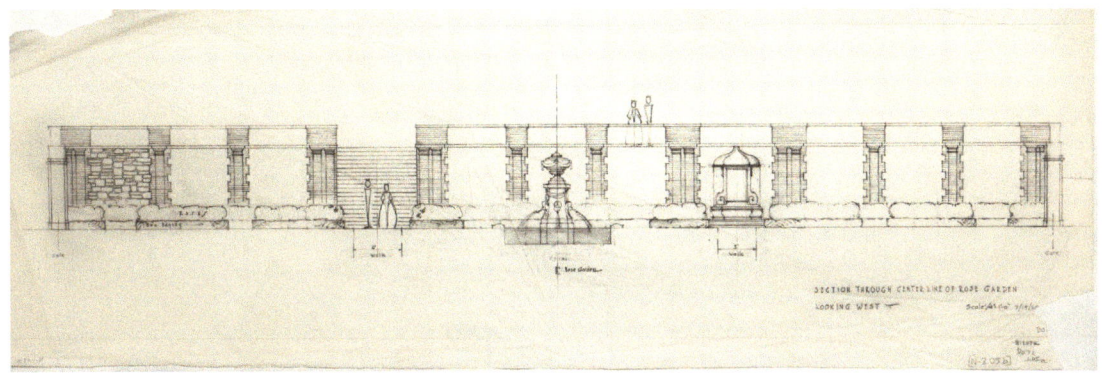

3.6
Section drawing showing the western wall of the Rose Garden with the Provencal fountain sketched in (it did not get placed there) and the reliquary for the Blisses shown inserted into the wall. Farrand adds figures for relative scale. Image from Dumbarton Oaks Garden Archives, Dumbarton Oaks Research Library and Collection.

3.7
Circular stairs from Green Terrace to the Swimming Pool. The use of watercolour emphasizes the romantic character of the stairs and the fountain, which was to be covered in moss. Image from Beatrix Jones Farrand Collection, 1866–1959, UC Berkeley, Environmental Design Archives.

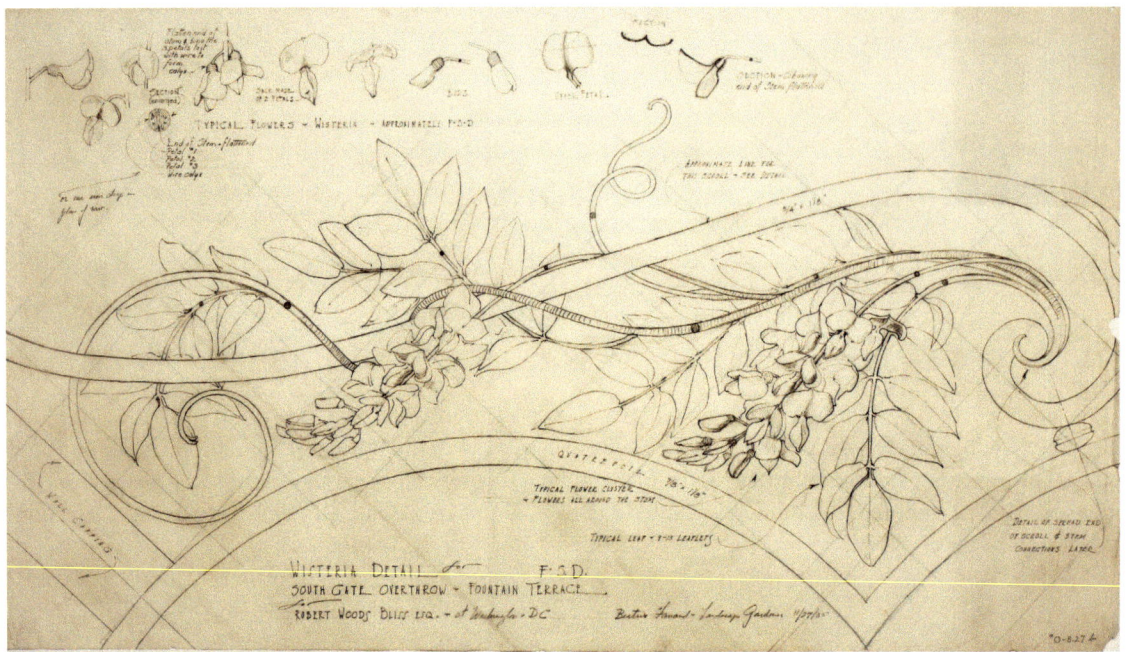

3.8
Full-Size Detail of Wisteria for south gate overthrow, Fountain Terrace. This drawing would be refined and then given to the ironworker to fabricate as a gate ornament. Image from Dumbarton Oaks Garden Archives, Dumbarton Oaks Research Library and Collection.

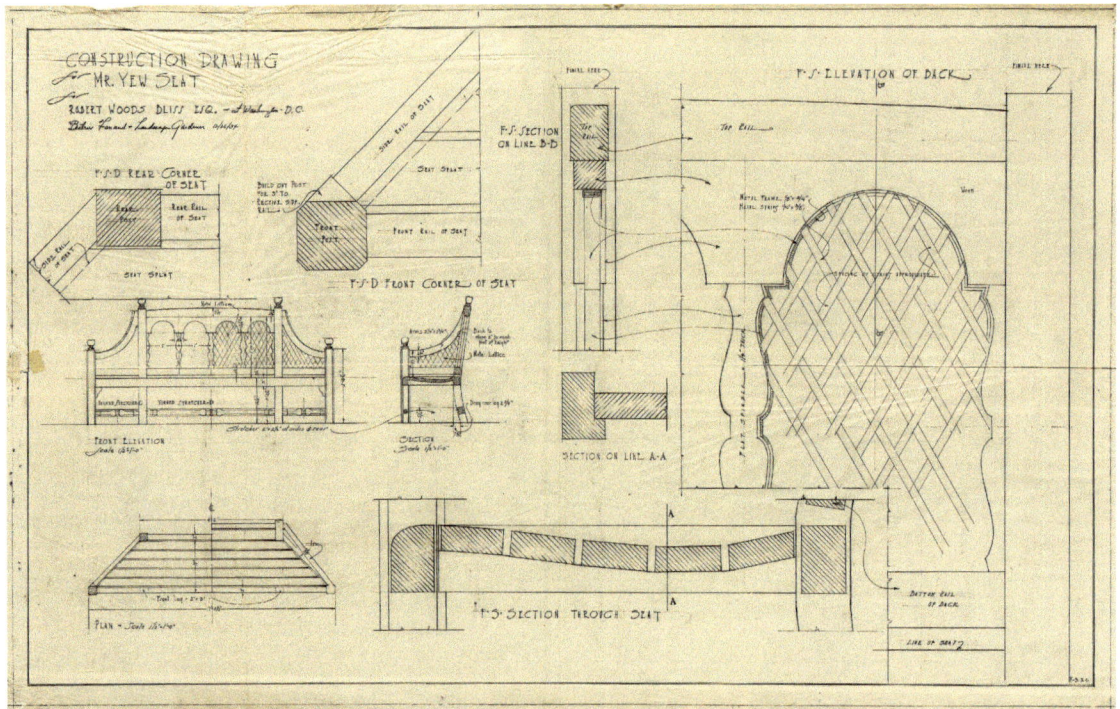

3.9
Construction drawing for Mr. Yew seat. Image from Dumbarton Oaks Garden Archives, Dumbarton Oaks Research Library and Collection.

up in place. It was at this level of detail that Farrand made notes on the drawings to adjust and refine the design. As she was often working with stone masons, iron forgers, and furniture makers, she would assure that the full-scale design was exactly what she desired before sending it to be fabricated. This process of considering the design at full scale allowed her to assess the smallest of details as well as the ways the materials might shape the final product. It was an attention to detail that is easy to miss until production; however, it required extensive time not only in the production of the drawing but also in the review and refinement process. A similar attention to detail was evident in the construction drawings for furniture, as Farrand designed each of the benches, tables, and chairs for the garden (Figure 3.9).

With the design concepts selected, Farrand would focus on carefully delineated plans that provided the details of the design, but more significantly established a clear design character of the garden room. This is evident in the drawing for the North Vista that includes the details of plantings, stairs, lawn, and walls, while also establishing the visual character of the garden room: a linear play of vision as the seemingly parallel lines merges at the northern edge (Figure 3.10).

Farrand's planting designs were created first as a concept, then as a drawing, and verified on-site for final decisions. For the flower and herb beds, she frequently began with plans that appear similar

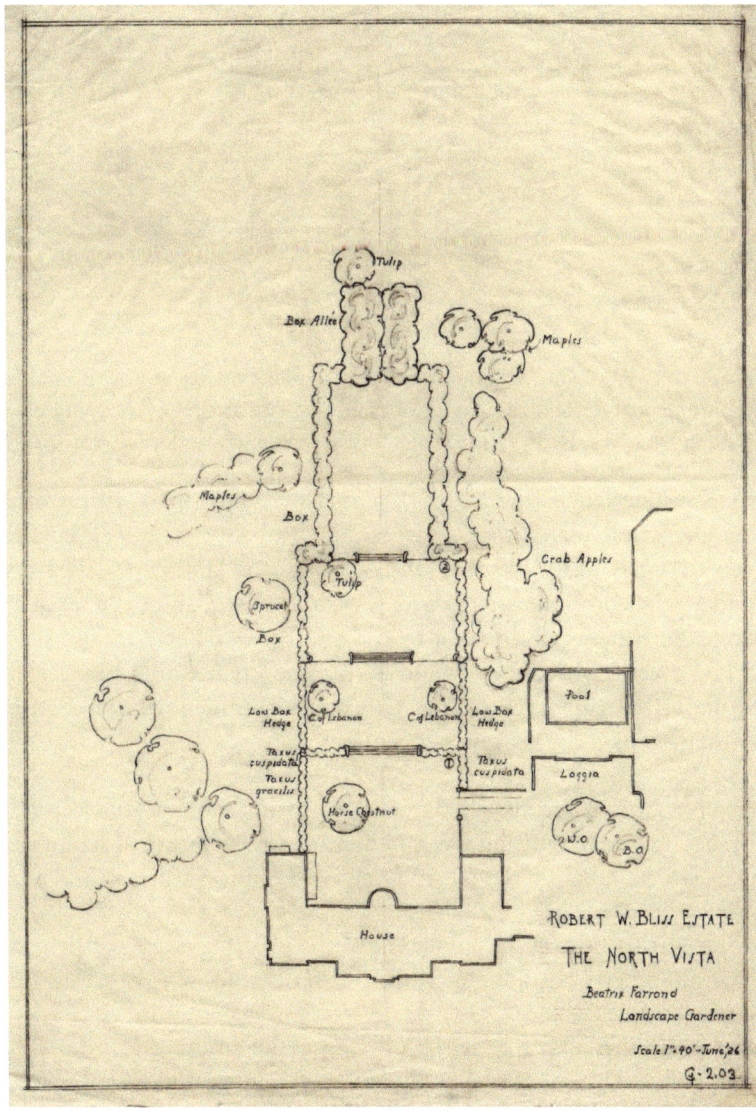

3.10
North Vista, June 1926, trace with graphite. The trees and the edges of the vista were most important for the effect Farrand was seeking, and thus, these are emphasized in the drawing. Image from Dumbarton Oaks Garden Archives, Dumbarton Oaks Research Library and Collection.

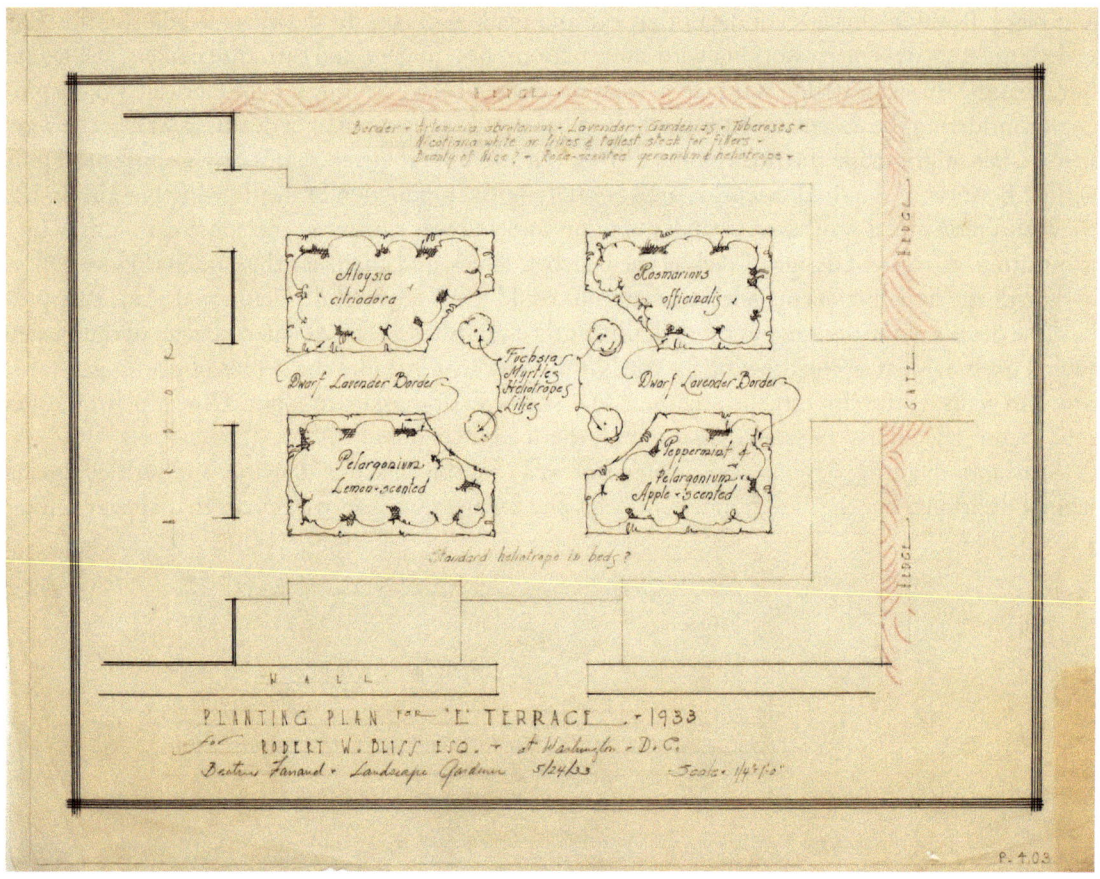

3.11
Planting plan for "E" Terrace, 1933, showing the initial simple planting design for this garden room. Farrand would determine the numbers and exact placement of plants on-site as they were being planted. Image from Dumbarton Oaks Garden Archives, Dumbarton Oaks Research Library and Collection.

in style to those by Gertrude Jekyll (Figure 3.11). Plants were identified within irregular spaces, all merging together as one whole plan. Evergreens, larger shrubs, or trees were identified by specific shapes such as circles, triangles, or squares. For larger gardens and landscapes, she frequently listed the plants to one side and placed corresponding numbers within the design next to each circle or irregular space identified as a plant. Rarely did she indicate the number of plants, nor did she seem concerned that they were marked exactly where they would be planted, as it would be decided at the time of planting depending on the individual specimens. This was the moment when a site visit was called for.

Robert Patterson described the process as one in which:

> She preferred to prepare a planting plan alone in her sitting room, a landscape clear on her inner eye, arranging her palette by writing plant names on a half bushel of white labels. Sorted into bundles, the labels were taken to the job and parcelled out to gardeners and assistants.[9]

She produced planting plans; however, actual details of the plantings were confirmed on the ground, thereby becoming a ground plan. Farrand described this process to Mrs. Bliss, suggesting that the

> exact position of [the] different plantations is difficult to determine academically, and the most important groups should be placed on the ground as alterations of grade, root space, and exact angle of vision should be the controlling factors, rather than their exact position upon a planting plan.[10]

Her hands-on approach was also a part of the construction and maintenance phases. As early as 1897, Farrand explained to a reporter, "I give orders to my men in person nearly always. I find it best to come in touch with them."[11]

A similar process might take place for the ornamentation or for a major structural addition to the garden. Where drawings did not provide the required information for a design detail, Farrand requested a full-sized mock-up that would be placed in the landscape for review. This approach was used for details such as the finials as well as for larger scale projects. Farrand's use of models at full scale underscores her belief that a drawing might not suffice for final decisions, but rather, a model that could be tested in the actual position and siting of the proposed element.

Farrand's designs for places such as Dumbarton Oaks have inspired designers and artists for over a century. However, her drawings, letters, and design process are less recognized. Farrand did not produce beautiful presentation drawings as she was focused on the land as a work of art rather than the drawings. Nevertheless, much can be learned from her approach in the combined power of using prose and drawing to develop ideas, with on-site refinements. As well, her attention to the development of a human scale in her gardens was developed through her drawings, for which every mark on the paper required a thoughtful intention, none generated by any other force than that of the designer. Her designs remain living works of art, produced through her exquisite use of prose, drawings, and on-site refinements. This is a process that remains powerful today as we seek to imagine, build, and steward places of inspiration for a better future.

Notes

1. Note that she was known as Beatrix Jones as this was prior to her marriage to Max Farrand. Jones, B. "Bridge over the Kent at Levens Hall." *Garden and Forest*, 1896, 9(412): 22; Jones, B. "The Garden in Relation to the House." *Garden and Forest*, 1897, 10(476): 132–133; Jones, B. "Nature's Landscape-Gardening in Maine." *Garden and Forest*, 1893, 6(289): 378–379.
2. Patterson, R. "Beatrix Farrand—1872–1959." *Landscape Architecture Magazine*, 1959, 49(4): 216.
3. Ibid.
4. Beatrix Farrand to Mildred Bliss, June 24–25, 1922. Garden Archives, Dumbarton Oaks Research Library and Collection, Washington, DC, p. 6.
5. Whitehill, W.M. *Dumbarton Oaks: The History of a Georgetown House and Garden 1800–1966*. Cambridge, MA: Harvard University Press, 1967, p. 67.
6. Mildred Bliss to Beatrix Farrand, July 13, 1922. Garden Archives, Dumbarton Oaks Research Library and Collection, Washington, DC.
7. Beatrix Farrand to Mildred Bliss, September 11, 1922. Garden Archives, Dumbarton Oaks Research Library and Collection, Washington, DC.
8. Ibid.
9. Patterson, *op. cit.*, p. 218.
10. Beatrix Farrand to Mildred Bliss, June 24–25, 1922, *op. cit.*, pp. 2–6.
11. "Miss Beatrix Jones's Vocation: She Does Landscape Gardening of All Kinds, from the Ground Up." *New York Sun*, October 31, 1897.

References

"Miss Beatrix Jones's Vocation: She Does Landscape Gardening of All Kinds, From the Ground Up." *New York Sun*, October 31, 1897.

Beatrix Farrand to Mildred Bliss, June 24–25, 1922. Garden Archives, Dumbarton Oaks Research Library and Collection, Washington, DC.

Beatrix Farrand to Mildred Bliss, September 11, 1922. Garden Archives, Dumbarton Oaks Research Library and Collection, Washington, DC.

Jones, B. "Bridge Over the Kent at Levens Hall." *Garden and Forest*, 9(412) (1896): 22.

Jones, B. "Nature's Landscape-Gardening in Maine." *Garden and Forest*, 6(289) (1893): 378–379.

Jones, B. "The Garden in Relation to the House." *Garden and Forest*, 10(476) (1897): 132–133.

Mildred Bliss to Beatrix Farrand, July 13, 1922. Garden Archives, Dumbarton Oaks Research Library and Collection, Washington, DC.

Patterson, R. "Beatrix Farrand—1872–1959." *Landscape Architecture Magazine*, 1959, 49(4): 216–218.

Whitehill, W.M. *Dumbarton Oaks: The History of a Georgetown House and Garden, 1800–1966*. Cambridge, MA: Harvard University Press, 1967.

4 Fletcher Steele, the Savvy Practitioner
Desire and the Cultivation of Connoisseurship

Martin J. Holland

Few landscape architects associated with the Grand Estate era were better known than Fletcher Steele (1885–1971). With close to 600 private commissions to his name, Steele eschewed most public projects, favouring working directly with an elite clientele. He cheekily observed that the fundamental problem with civic work is that "everything good is flattened by committees."[1] However, it was not just the preservation of the "good" that guided his focus on private residential work for Steele. The formation, and careful cultivation, of personal relationships with his clients is what ultimately sustained him and his practice.

Known to his clients for his rapier wit and impeccable taste, Steele easily navigated social circles well above his middle-class origins in Rochester, New York. Attending Williams College from 1903 to 1907, Steele then attended Harvard's recently formed graduate program in Landscape Architecture.[2] He never completed the curriculum required for matriculation because he was lured away from his studies by an offer of mentorship from Warren H. Manning, a founding member of the American Society of Landscape Architects (ASLA). Although not an offer of paid employment, the opportunity gave Steele unfettered access to Manning and his design methodologies. In a letter to his mother, Steele remarked that his first day with Manning provided welcome relief to his studies at Harvard: "I learned more about landscape proper than I did all last winter."[3] He found the work engaging and challenging as compared to the academic exercises replicating the Beaux-Arts style. Robin Karson, in her definitive treatment of Steele, observed that "unlike many other students, Steele never learned to draw freehand and spent considerable time mastering the rudiments that his instructors considered fundamental to a well-rounded education in the field."[4] The issue of graphical ability may have been a sore point for Steele, who in another letter to his mother justified his decision to set aside his studies, by emphasizing that he was not a draftsman: "I am really an assistant directly to Mr. Manning and as such I get an insight and an oversight into the whole working profession that I could not get in any other way in the world."[5]

Manning was indeed providing him with the technical expertise, but more importantly, the social networks necessary to succeed within the profession, although Steele was growing more aware of the limitations of his mentor's focus on "…convenience and economy and natural conditions, whence he [Manning] believes beauty will follow."[6] Steele considered Manning to be more of an engineer, while the landscape architect was for Steele more akin to an artist. The profession required understanding the roles of composition, colour, texture, and form in their respective applications. This artistic ideology only became more refined and entrenched in Steele's beliefs, articulated in his 1934 article entitled "Fine Art in Landscape Architecture," where Steele asserted, "Beauty exists not in appeal to all the senses only, but to the mind as well, and probably the unconscious itself. Beauty is not obvious until it has been shown us by the artist." He goes on in that article to compliment landscape architecture as "the most pretentious of the plastic arts. It deals with landscape elements of the utmost grandeur that can be moved by man. All the world is peculiarly his realm with its infinite possibility of beauty-relationships."[7] Manning was willing to partially sponsor Steele's first "Grand Tour" of

DOI: 10.4324/9781003183402-4

Europe and North Africa, providing him with four months away from the office, a hundred dollars to help defer the costs of travel, and the use of office staff upon his return to edit and formalize his report.[8] Steele ultimately left Manning's tutelage in 1913 to start his design practice. Steele developed his extended network of clients through letters of introduction, personal invitations to social events, and providing engaging public lectures.

But it was through his popular writings that Fletcher Steele began to cultivate his reputation as a connoisseur. Eventually authoring two books—*Design in the Little Garden* (1924) and *Gardens and People* (1964)—it was his numerous articles appearing in such varied publication venues as *Ladies Home Journal*, *House Beautiful*, *Garden Magazine*, *Horticulture*, *Landscape Architecture*, and even *Vogue*, that firmly established him as a professional within high society. He became the darling of an elite class of wealthy women and was, as Karson describes him, "the most faithful male member" of The Garden Club of America by 1920.[9] Steele expressed a sympathetic view of the class situation of his female clients:

> In business, in charity, clubs and family life, this person is tied down in a thousand ways necessary to decent communal life. Those obligations have become habitual until they have come to believe that their real and genuine selves [are] bound up in living for others and with other people considerately. If that were so, in fact, they would not be good wives and mothers. They would be nuns. Every human being keeps to the end a personality which is private, with its own private desires and satisfactions often quite unrelated to the everyday life they lead, which indeed, they do not want to change. These unspoken desires and satisfactions come to life in daydreams. Daydreams hurt nobody.[10]

Attuned to their needs, Steele was also able to instil a taste for experimentation within his clients. As Karson has argued, Steele successfully supported and reinforced his clients' desires to "play at being someone else, somewhere else."[11]

A telling example of this is Steele's design for a swimming pool for Dr. John Washburn Bartol, a prominent, Harvard-trained physician who served as the president of the Massachusetts Medical Society, the Boston Medical Library, and the Boston Health League. The son of a Unitarian minister and the father to four daughters, Bartol was also Chairman of the Board of Simmons College, a women's college in Boston (where a building was named in his honour in 1953).[12]

While not likely rendered in Steele's hand, the drawing (Figure 4.1) is probably the artistry of Henry Hoover, who started working for Steele on a part-time basis in 1925 while Hoover was still completing his architectural education at Harvard. The drawing is richly illustrated in painstaking detail and strictly adheres to scale, offering multiple textures and line weights within the monochromatic rendering. While plant species are not indicated, a close inspection of the drawing allows distinction between evergreen and deciduous, while quick yet precise cross-hatching indicates variations in groundcover. Specific locations are identified and labelled, providing enough context about their purpose and allowing the viewer to imagine themselves and their loved ones inhabiting the design.

While this drawing is a considerable departure from the garden designs for which Steele had built a formidable reputation, the swimming pool illustration is notable for a host of reasons. The form of the pool is a departure from the typical typology, where there is not a right angle to be found within the perimeter. Instead, the careful pencil rendering inserts rock formations that occupy the southwest corner disguising the pool's southern wall with a stone outcrop that serves as the elevational change for a slide and high diving board, complete with stairs to access these amenities. (For an alternative vignette of this design, please examine Figure 4.2). The mortared fieldstone southern wall slides into the constructed outcropping as the terminus. The western edge of the pool appears utilitarian, with grass or turf nearly meeting the water's edge, save for the presence of a stone coping providing a secure footing as one dives into or exits the water. The north edge of the pool is defined by an informal, almost wandering line of large fieldstones, heading north by northeast, broken in places with reeds, grasses, and other hydrophilic vegetation. Also present within the area is the "low board" for those either too timid for the higher board or those who wish for a quick exit and immediate re-entry

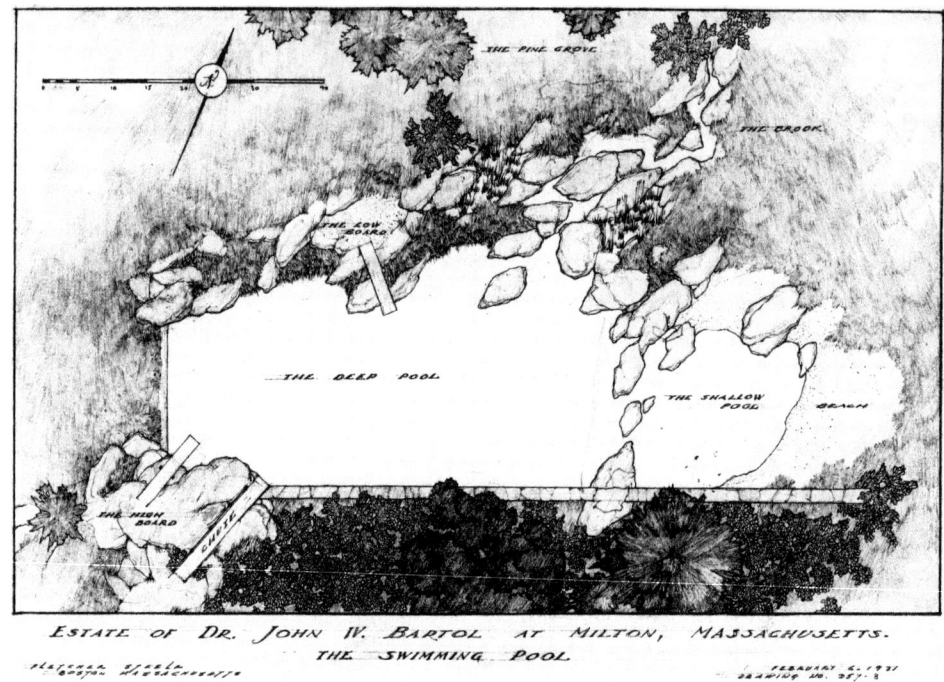

4.1
The Swimming Pool (Drawing No. 357-8). This image is from the collection at the Terence J. Hoverter College Archives, F. Franklin Moon Library, State University of New York College of Environmental Science and Forestry, Syracuse, NY.

into sparkling water. Further to the northeast, the pool appears to be fed by a babbling brook just a few feet to the north of the perimeter of the pool, providing the artifice that this swimming pool is a constructed refinement of the naturally occurring, and quintessentially American, "swimming hole."

Further to the north, a pine grove beckons to the adventuresome. A line of partially submerged stones further delimits the deep pool from the shallow, although the presence of four gentle strokes of a hard pencil infers some form of elevational break, possibly implying a short cascade or waterfall that separates the two pools. This would require substantial hydrological engineering and equipment that would have been out of technological reach given the era of the design. (In Figure 4.3, it is clear that the water level is consistent across the entire surface of the pool and that there is also a revision to the drawing, where the stones are placed as a demarcation between the deep and shallow area). The presence of sand, stippled and shaded, offers an ideal playground to children who are either unsure swimmers or need a break from the cold, refreshing water. Large stones also provide places of repose, allowing children and parents alike to bask in the warmth of the sun. Feathery grasses meet the beach's organic edge, providing and defining the small pocket of sand magically appearing from the rich agricultural land of the surrounding property (Figure 4.4). The heavily planted southern edge offers both privacy and shade, while the pool's coping incorporates an emerging boulder that anchors the demarcation between the deep and shallow pools.

What is remarkable about this pool design in the early 1930s is how it is simultaneously modern and fundamentally nostalgic. The fieldstone pool wall suggests the stone constructs that separated and defined agrarian areas and helped define land ownership. Yet, it is an unusual choice of materials for a swimming pool coping, given the potential for cracking and mortar failure. Yet, the design references well-known cultural typologies (that of the swimming hole and the beach—both firmly associated to New England audiences with notions of leisure) while providing something entirely new.

The vignette labelled "Diving Tower and Chute" (Figure 4.2) further reveals this tension between nostalgia and the modern, and again, was most likely the product of Henry Hoover's talented hand. The illustration depicts the southwest corner of the swimming pool, complete with a cast in place concrete tower and a slide carefully nestled between rock outcroppings. The tower is particularly

4.2
Diving Tower and Chute (Drawing No. 357-11). This image is from the collection at the Terence J. Hoverter College Archives, F. Franklin Moon Library, State University of New York College of Environmental Science and Forestry, Syracuse, NY.

noteworthy as the spiral stairs leading upwards are all dependent upon internal cantilevered rebar or ironwork for their stability and strength. No railing is present along the stairway save for a simple rope along the interior circumference of the tower itself. The platform situated on top of the concrete shaft is semi-circular, which further suggests cantilevered supports for the surface of the platform. The tower has openings within the ropework at the rear and front, allowing for access from the spiral staircase and, of course, to the pool below. A darker area of water to the lower left-hand side of the drawing draws a viewer into this constructed reality, as the client imagines him or herself leaping from the tower or swiftly descending from the water chute into the deep water. The plantings adjacent to the chute are significant and diverse, with the drawing relying on the tonal variations of light and dark to give shape to the canopies and branching structure.

While no direct correspondence between Fletcher Steele and John Bartol could be located discussing the design proposal illustrated in Figure 4.2 (Drawing No. 357.11), it was nonetheless revised with a rendering that hints at a return to a simpler time. The drawing in Figure 4.5 (Drawing No. 357-20-A) eschews the use of a diving tower in deference to simple diving boards secured to the significant

4.3
Three figures at the pool. This image is from the collection at the Terence J. Hoverter College Archives, F. Franklin Moon Library, State University of New York College of Environmental Science and Forestry, Syracuse, NY.

4.4
Existing site conditions before construction of the pool. This image is from the collection at the Terence J. Hoverter College Archives, F. Franklin Moon Library, State University of New York College of Environmental Science and Forestry, Syracuse, NY.

4.5
Detail of the Swimming Pool (Drawing No. 357-20-A). This image is from the collection at the Terence J. Hoverter College Archives, F. Franklin Moon Library, State University of New York College of Environmental Science and Forestry, Syracuse, NY.

elevation above the swimming pool's surface. While the original plan drawing of the pool possessed a clearly defined stone stairway that led to the diving boards, he simplified the design by including a rustic wooden ladder resting against the outcropping as if it were borrowed from a nearby barn. The drawing further reveals the "natural" setting of the pool, complete with emergent vegetation finding the available soil within the nooks and crannies of the stone to take root, with the illustration providing a clear example of how the fieldstone wall would terminate against the larger stone outcropping. The season represented in the drawing is undefined; the tree dominant in the background to the left is leafless either representing the seasonality of the drawing itself (drawn in October 1931) or perhaps in the early spring before it has had a chance to leaf out. Branches to the upper right corner indicate either persistent leaves or indicate that early spring is occurring. Records from Steele's office indicate that two diving boards were indeed purchased on November 12, 1931, from the Beacon Equipment Company in Brookline, Massachusetts (at the price of $62.50 each), showing that the latter design ultimately was selected by the client.

This plan (Figure 4.6) includes the vital topographical information necessary for the proper siting of the swimming pool including numerous spot elevations located on a nearby hillock to the northeast. The drawing offers the most specificity around the immediate environs of the swimming

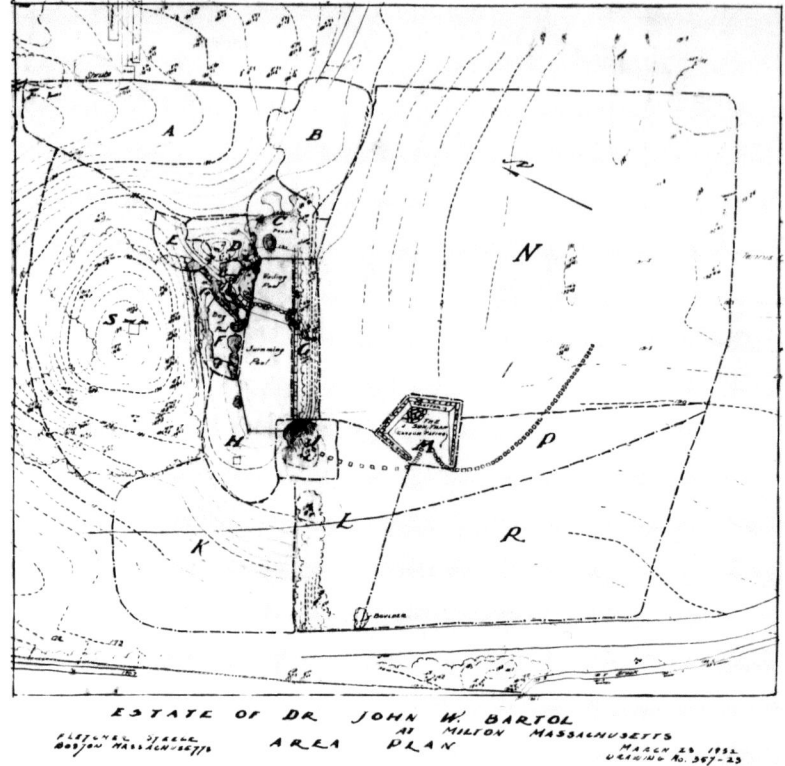

4.6
Area Plan (Drawing No. 357-23). This image is from the collection at the Terence J. Hoverter College Archives, F. Franklin Moon Library, State University of New York College of Environmental Science and Forestry, Syracuse, NY.

pool and now uses the language of the activity itself (swimming vs. wading) as distinguishing characteristics. Also, note that there is no scale indicated within the drawing. The area plan provided to John Bartol was a typical strategy for Steele in the early 1930s, leveraging a single design commission to lay the groundwork for possible future projects.[13] Care has been taken to subdivide the property into several distinct areas that Steele most likely identified as future potential projects. Signified by a corresponding letter of the alphabet, this technique allowed his clients to select the next possible area for design in a piecemeal manner, giving voice to their desires while keeping the projects discreet and manageable.

In a lecture in Westport, Connecticut, Steele remarked that the landscape architect must refuse "to be turned out in the front yard like a dog while his client holds him on a leash demanding 'what would you do here?' [or] 'Make me a pool under that tree.'" He goes on to advise that the savvy designer will instead wander into the client's house to see what kinds of books she reads and what kinds of furniture and bric-a-brac she gathers about her. He gets her talking about her travels and the places she likes best and the ones she does not like. He probes to discover, not what she has, but what she dreams of having; not what she does but what she would like to do.[14]

Although unusual for its time, Steele's design for the Bartols' swimming pool was aesthetically appropriate for their "farm," the vacation home outside of Boston. Executed through Hoover's fine rendering, the design delivered on the clients' attachment to Americana, including the idea of a found "swimming hole" with a mythic water source. Where land had been cleared and despite hydrological realities, now "the pine grove," "the beach," and "the brook" appear as naturally occurring elements in their idealized form. Daydreams of the new and desires for a seemingly familiar past are simultaneously fulfilled. That Steele's design is closely aligned with his client's desires is exemplified by the gravestone of Mrs. Bartol, which reads, "Briskly walked with thoughts of cheerful yesterdays and confident tomorrows."[15]

Notes

1 Karson, R.S. *Fletcher Steele, Landscape Architect: An Account of the Gardenmaker's Life, 1885–1971*. Amherst, MA: Library of American Landscape History, 2003, p. xii.
2 Ibid., p. 7.
3 Ibid., p. 9.
4 Ibid., p. 8.
5 Ibid., p. 11.
6 Ibid., p. 12.
7 Steele, F. "Fine Art in Landscape Architecture." *Landscape Architecture*, 1934, 24(4): 178. June 21, 2021. http://www.jstor.org/stable/44661215.
8 Karson, *op. cit.*, p. 12.
9 Ibid., p. 47.
10 Ibid., p. 103.
11 Ibid., p. 46.
12 Obituary found at https://www.newspapers.com/clip/54083794/obituary-for-john-washhurn-bartol; Information regarding building dedication, https://www.simmons.edu/library/archives/history-exhibits/brief-history-simmons, both accessed June 23, 2021.
13 For another example of this strategy, please refer to Karson, *op. cit.*, p. 85.
14 Ibid., p. 103.
15 See website, https://www.findagrave.com/memorial/142730450/charlotte-hemingway-bartol, June 23, 2021.

References

Karson, R.S. *Fletcher Steele, Landscape Architect: An Account of the Gardenmaker's Life, 1885–1971*. Amherst, MA: Library of American Landscape History, 2003.

Steele, F. "Fine Art in Landscape Architecture." *Landscape Architecture*, 24(4) (1934): 178. http://www.jstor.org/stable/44661215.

5 Topographical and Landform Explorations

Revisiting Noguchi's Sculptured Landscapes and Their Representations

Shannon Bassett

Within the context of the resurgence of a preoccupation with constructed topographies and landform explorations in critical contemporary landscape design—and, by extension, critical architectural design taking its cues from landscape—it is instructive to look back at the seminal landscape projects of Isamu Noguchi. The following investigation explores the materials and methods of representation of these landscapes' sculpted topographies and landforms. This includes his generative sculptures and study models, made predominantly of clay, and their iterations into inhabitable landscape and architecture. The projects discussed here include both his realized constructed projects and speculative projects that informed his realized projects.

Critical contemporary practice engaged with topographical and landform design has also been layered with performative ecological designs responding to our current pressing environmental challenges. This has included grading design which curates hydrological flows including swales and rain gardens. Critical landscape practices such as STOSS and Chris Reed are actively using topographical explorations as performative ecologies in order to curate effect such as in their Bass River Park in Massachusetts. Here, sculpted topographies choreograph the collection and treatment of water, while responding to changing environmental conditions. Michael Van Valkenburgh's recent Corktown Commons, located in Toronto on the banks of the Don River, is a 16-foot-high flood protection landform and urban park. The site and its newly constructed topography were previously a flat, barren brownfield. Adriaan Geuze and West 8s articulated and sculpted urban wave decks both puncture and pontificate several former spits along Toronto's waterfront, at the termination of Toronto's Cartesian street grid.

In critical contemporary architectural design—working at the intersection of landscape or blurring the boundaries between architecture and landscape—strategies of sculpted and constructed landscapes include seminal projects such as Diller Renfro Scofidio's extension to the existing Plaza at Lincoln Centre in New York City into a folded landscape and new inhabitable lawn. This formal constructed landscape has new programs tucked underneath it, including a restaurant. Infrastructurally, it also serves as a circulation connector from the sidewalk and street level to the parking located beneath the elevated plinth of the original plaza. Maritime Youth House by PLOT Architects—Bjarke Ingels and Julian De Smedt—plays with sculpting the landscapes of a wooding decking surface as a topography, with programs and spaces carved out and tucked beneath this new topography or ground. At a larger earthwork and infrastructural scale, projects such as Weiss Manfredi's Seattle Sculpture Park operate as infrastructure and landscape. A land ramp connects Seattle's downtown street level to its waterfront, spanning over several layers of infrastructure, including rail and a highway. It also operates as an open-air sculpture park and circulatory promenade.

DOI: 10.4324/9781003183402-5

While teaching landscape design studios and architectural studios engaged with landscape, overarchingly, design students are not using hand models anymore as initial generators for spatial concepts and form. Hand models working synthetically in analogue have been replaced by digital generators and digital representation. This precludes the richness and depth of the direct hand-eye cognitive exercises of sculpting with material. Often, students are engaged with parametric modelling and algorithmic design through such modelling programs as Grasshopper. These are generative of complex geometrical landscapes or folded landscapes. What are the possibilities afforded to the design student by revisiting an iterative design process grounded in using as study models more primal and malleable materials such as modelling clay? There are further potentials that this analogue process of working as part of the iterative process can be hybridized with digital, as the sculpted clay models as generators can then be translated into other generative forms such as laser cutting and 3D printing.

At the same time, there is also a value in revisiting the now status-quo conventional representation of topographical variation and shifting landscape through the use of the more two-dimensional representation of so-called contours, which are laser-cut. These are, in fact, an abstracted representation where often built-up layers and laminations of cardboard represent topographical and site elevational change. However, these abstractions of the actual landscape are not truly representative of a "sculpted landscape." When used as study models meant to be generative, their "fixedness" and two-dimensionality pre-empt the more open-ended and looser spatial and conceptual explorations of form and ensuing programs generated from topography or landform. Traditional site materials such as clay are malleable, lending themselves amenability to quick generative volumetric studies, which can be more improvisational and modifiable in several iterations. Clay can be scored linearly, in addition to being kneaded, pinched, or dimpled. Clay site models can also be understood at multiple scales, in part or the composition of their parts to the whole.[1]

Isamu Noguchi's landscape project designs discussed here were predominantly designed through hand modelling with sculpted clay. Noguchi's initial spatial exploration through sculpture and clay was improvisational and conceptually generative. These study models might be further translated into more "fixed" and static materials such as cast plaster or bronze. Noguchi's unrealized and speculative projects discussed here were of those materials. Projects such as the Riverside Playground in New York City did have drawings generated from translations of Noguchi's earlier study models, as it was nearly implemented as a realized project. The landscapes which were built were translated into inhabitable living landscapes and sculptured earthworks. They used earth and living materials such as vegetation, tree cover, and moss to cover their earthworks. Sculptures at 1:1 scale also pre-figured in the landscape designs. Some were translated into human scales, which served as furniture generated from notions of ergonomic topographies. Selected stones were also spatially deployed, with attention to solid and void and the spaces between. Surfaces were choreographed with hardscapes such as raked stone.

The projects discussed here are either constructed or speculative. Interestingly, as the landscape and architectural historian Marc Treib has noted:

> Noguchi rarely made sketches and worked mostly—and sporadically—in study models, and changed a lot on site (in the field). The construction drawings (and even many of the study drawings), were done by the architects or by Shoji Sadao, who worked with Noguchi (as well as being a partner to Buckminster Fuller).[2]

Also worthy of revisiting are the programmatic propositions of Noguchi's series of playground designs, embodying children's play which was to be non-directive, explorative, and open-ended, in opposition to the more prescriptive children's playground design ensued at the end of the 1970s and into the 1980s. Further, within the context of landscape or ecological urbanism, a revisiting of Noguchi's topographical strategies and their latent urban potentials is instructive.

In his seminal text, "Programming the Urban Surface," Alex Wall states that "there has been a renewed interest in the instrumentality of design and its enabling function-as opposed to representation and stylization."[3] As he points out, "landscape no longer refers to the prospects of pastoral innocence, but rather invokes the functioning matrix of connective tissue that organizes not only objects and space, but also the dynamic processes and events that move through them." Landscape, "as an active surface, structure[s] the conditions for new relationships and interaction among the things it supports," becomes a particularly useful model for architecture students when beginning to read and gain an understanding of the landscape of a given site condition. Traditionally defined "as the art of organizing horizontal surfaces," landscape is characterized by "a strategic deployment of processes, both natural and man-made, as well as being a medium which is capable of temporal change."[4]

UNESCO Garden: Garden of Peace, Paris, 1958

Noguchi's sculpted topographies and landscapes existed both as two-dimensional reliefs or terrain as well as inhabitable three-dimensional topographies. Most notable was Noguchi's UNESCO Garden in Paris, for the UNESCO House, which operated as the former. The overall design comprised an upper delegates' patio and terrace, which was 32 m × 32 m. The terrace was adjacent to the lobby and linked with the architecture. There is also a lower garden with more contemplative spaces. Here, Noguchi fused his sculptural reliefs into a habitable relief. The forms, as experienced in their entirety and before the vegetative cover, came in. The landscape worked with architect Marcel Breuer's brutalist concrete architectural design for the UNESCO House, elevated on 'pilotis.' The design of the landscape terrain could be experienced from multiple scales. Experienced from the adjacent balconies "from above," the topography could be experienced in its entirety as an aerial—similar as to how it would be experienced as a clay model in Noguchi's studio—which revealed all of the sculpted berms or mounds of the design. As experienced at the human, experiential scale at ground level, the sculpted topography created rich spatial experiences. It curated circulation and flow, with spatial design and gestures borrowed from traditional Japanese stroll gardens. The mounds and the movement through act more at the scale of bonseki in certain places.

Noguchi played between scales of his translated initial sculptural studies. He incorporated a more detailed 1:1 sculpture in the form of a large stele set in the pool of the upper terrace, which operated as a vertical element in the garden. In this design, the sculptural forms, mounds, and biophilic forms create volumes for the storage and collection of water. Noguchi also played with hardscape of raked gravel and softscape including a soft moss that worked on his sculptured mounds and softened their hard geometries. His furniture design is also sculptural and a scaled topography shaped to ergonomics and works within the flat plane of the garden design.

Reader's Digest Building, Tokyo, 1951

As does the UNESCO Garden, the landscape design for the Reader's Digest Building in Tokyo operates as a two-dimensional relief or terrain. The project further iterates the potentialities of the sculpted topographies and landform to being curating a hydrological cycle. The sculptural and topography include more pronounced watercourses, ridges, and hills. The forms are sculpted and also performative in that they deliberately choreograph and curate the movement and flow through the garden space.

Noguchi's Playgrounds

Noguchi's series of playground designs pushed the notion of the inhabitation of his sculpture into inhabiting the spaces of the sculpture itself three-dimensionally, as opposed to being a two-dimensional relief which was experienced more from above and moving around as described in the UNESCO Garden in Paris and the Reader's Digest landscape design.

The playground designs could be inhabited from the inside. Their programs are open-ended and overlaid with the program of play and non-determined play. These start from a process of sculptured clay that iterated into the more pronounced spatial inhabitation and scale. The series of designs he did for playgrounds engaged in concepts of "non-directive" play arguably generated from Noguchi's open-ended and loose use of clay as his design.

Riverside Park Playground, New York City

Noguchi's proposal for the Riverside Park playground was in collaboration with the architect Louis Kahn. Iterative study models were more abstractions of sculpted landscapes; their subsequent iterations into an inhabitable playground for children became more architecturally scaled. There was also a translation into more measured architectural construction drawings. We began to read what might be a retaining wall to the edge of the site bordered by road infrastructure, which also uses topography sectionally step down and create a visual, experiential boundary and separation from the road. The sculpted volumes become translated into scaled ergonomics for children's play. Noguchi's scored surfaces become surface textures of hardscapes which create open-ended programs. Unfortunately, the Riverside Park playground landscapes were not realized.

Play Mountain

Noguchi's design for Play Mountain in 1933 was the first project where he translated his more abstract relief sculptures into a contoured spatial relief and inhabitable project into the form of a proposal for a playground. It was an actual inhabitable playground, working between being a public sculpture and a landform. It was activated with the program of a playground for children. He proposed this when Robert Moses was the New York City Parks Commissioner at that time; however, the proposal was rejected. Noguchi had several other playground proposals which were also dismissed for New York City.

As cited by Noguchi, he

> wished to bring [his] vision of a pyramid in Idaho into the experience of people in the city where [he] lived. Play Mountain was conceived as a way of building on a city block in New York a mountain, which would, in fact, be an enhanced area for children's play. There were steps of all sizes, a slide with water in the summer, a longer one for sledding in winter. These sloping surfaces could also serve as a roof.[5]

Realized Playground: Moerenuma Park, Japan, 1998–2005

As has been noted in the literature, playground design changed at the end of the 1970s. The one playground that Noguchi had realized, at Moerenuma Park, was a compilation of his previous sculptural design explorations in playground design. The architect was Shoji Sadou. The design is folded into a bend in the river. Again, as the landscape and architectural historian Marc Treib noted, it was Sadou translating the more abstracted conceptual design Noguchi generated in clay and sculpture into the more "architectural" translations of construction drawings, which took into consideration grading, scale, building codes and conventions, and more ergonomic considerations of the user.

Conclusion

In summary, there are many opportunities afforded by revisiting and re-reading the topographical and landform explorations in Noguchi's landscape designs. The sculptural forms generated by an open-ended, more improvisational spatial study with clay hand models afford new design possibilities and reconsiderations for our current pressing environmental challenges such as hydrological design and performative ecologies.

Further, a return to the more direct cognitive hand-eye creative function in the sculpting of material offers more generative design forms for both landscape design students and professionals, which can at once be further generative when combined with as part of a more extensive hybrid analogue and digital process of working. Figures 5.1–5.11 showcase a collection of Noguchi's models and drawings.

5.1
Looking out over UNESCO Garden from the balcony of Marcel Breuer Building, Paris. The Noguchi Museum Archives, 00000. © INFGM/ARS.

5.2
UNESCO Garden, Lower Garden at bottom and Delegates Balcony above. The Noguchi Museum Archives, 00000. © INFGM/ARS.

Topographical and Landform Explorations

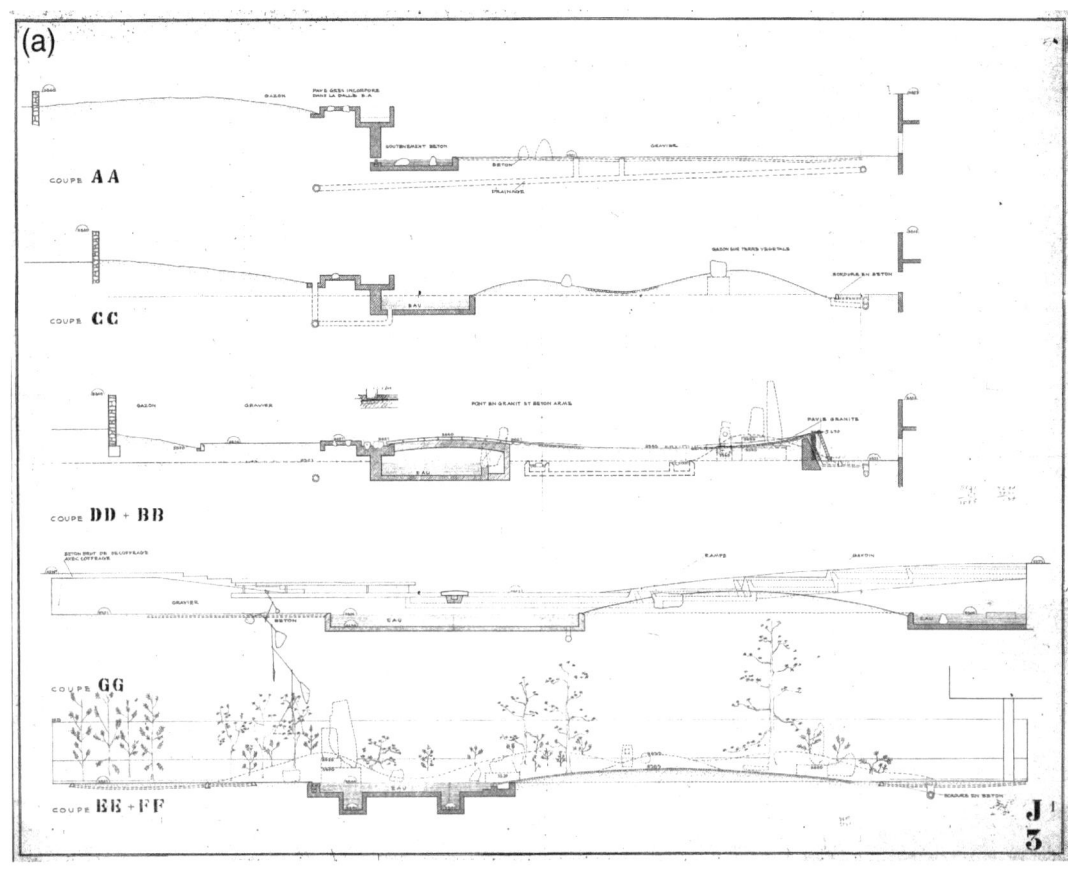

5.3
(a) Construction Drawings for UNESCO Garden. The Noguchi Museum Archives, 00000. © INFGM/ARS.
(b) Readers Digest Landscape, Tokyo, Water Course construction. The Noguchi Museum Archives, 00000. © INFGM/ARS.

5.4
Child with Model for Riverside Playground (Proposed) in Noguchi's Studio. The Noguchi Museum Archives, 00000. © INFGM/ARS.

5.5
Riverside Park Playground Model. The Noguchi Museum Archives, 00000. © INFGM/ARS.

Topographical and Landform Explorations

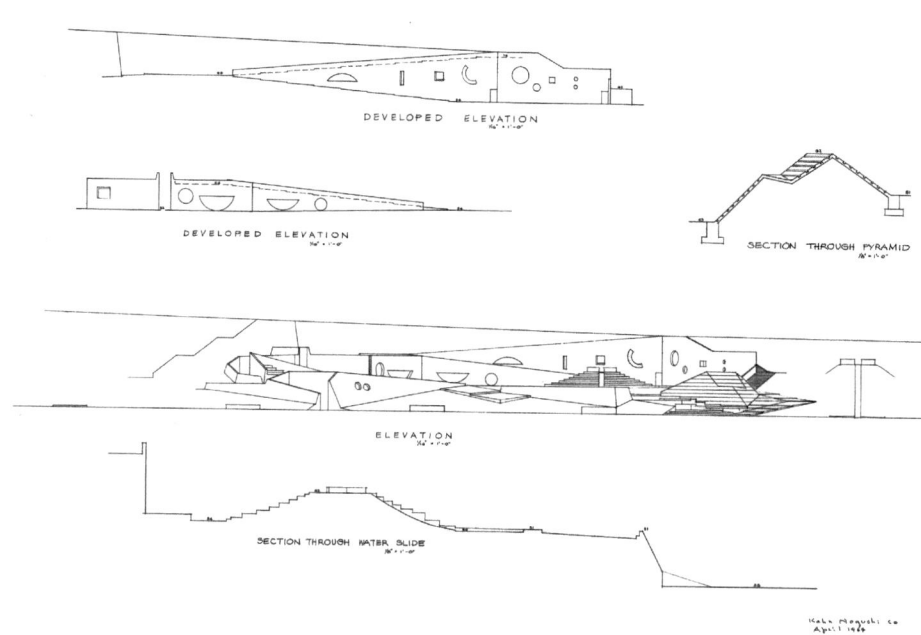

5.6
Riverside Park Playground Drawings. The Noguchi Museum Archives, 00000. © INFGM/ARS.

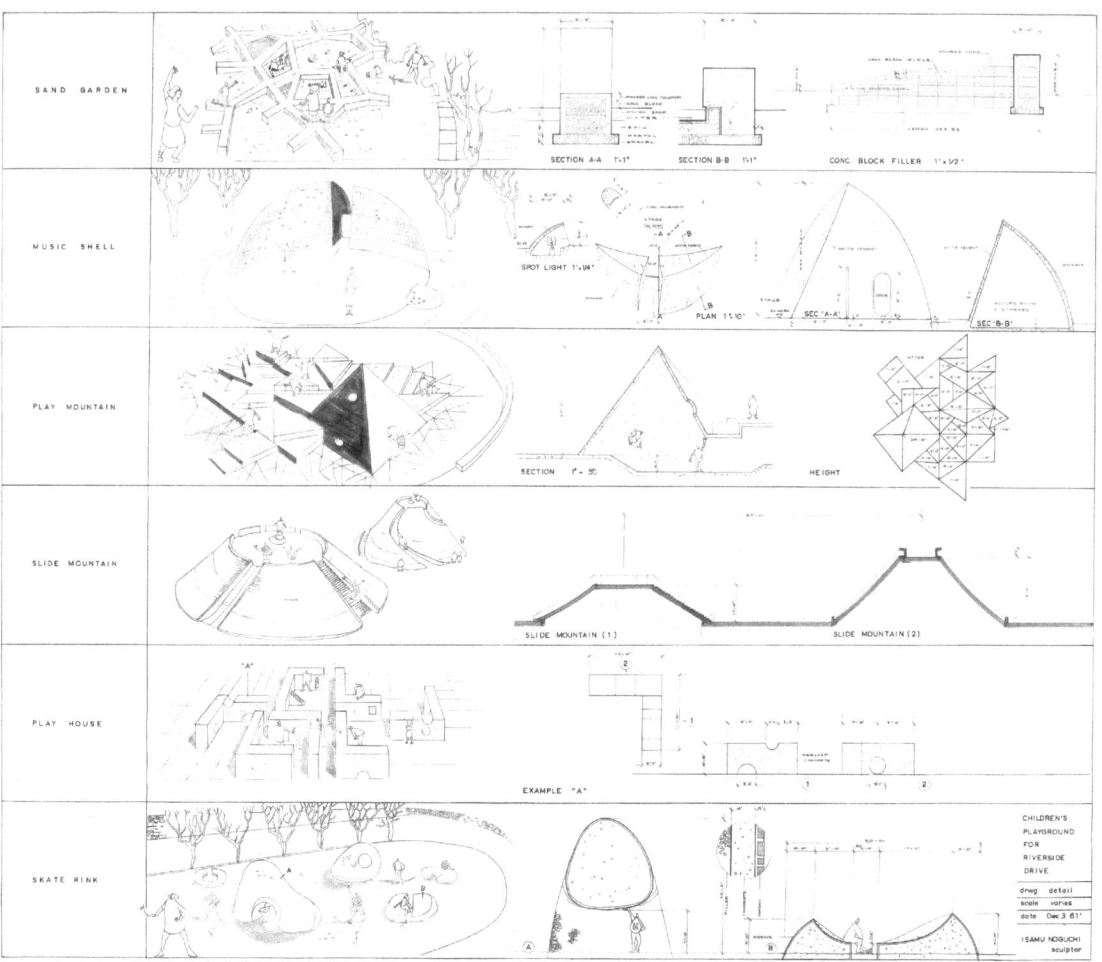

5.7
Riverside Park Playground Detail Drawings. The Noguchi Museum Archives, 00000. © INFGM/ARS.

5.8
United Nations Playground, New York City. The Noguchi Museum Archives, 00000. © INFGM/ARS.

5.9
Play Mountain, Plaster Model. The Noguchi Museum Archives, 00000. © INFGM/ARS.

Topographical and Landform Explorations 41

5.10
Moerenuma Park earthwork. The Noguchi Museum Archives, 00000. © INFGM/ARS.

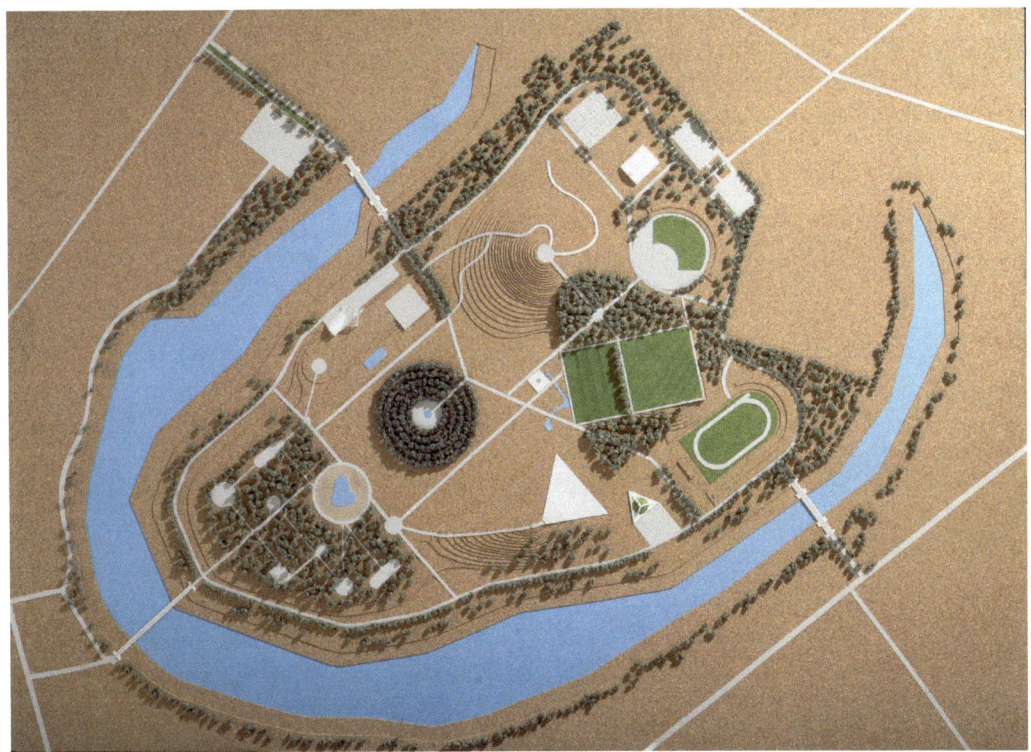

5.11
Moerenuma Park model of site design. The Noguchi Museum Archives, 00000. © INFGM/ARS.

Notes

1. Rieder, K. "Modeling, Physical and Virtual." In: M. Treib (Ed.). *Representing Landscape Architecture*. New York, NY: Taylor & Francis, 2008, pp. 168–187.
2. Email conversation between the author and Marc Treib, April 10, 2021.
3. Wall, A. "Programming the Urban Surface." In: J. Corner (Ed.). *Recovering Landscape: Essays in Contemporary Landscape Architecture*. New York, NY: Princeton Architectural Press, 1999, p. 233.
4. Waldheim, C. "Landscape Urbanism: A Genealogy." *PRAXIS: A Journal of Writing + Building*, 2002, 4 (Landscapes): 12.
5. From Noguchi Archives.

References

Lee, U. (ed.). *STOSSLU*. Seoul, South Korea: C3 Publishers of Korea, 2007.
Rieder, K. "Modeling, Physical and Virtual." In: M. Treib (Ed.). *Representing Landscape Architecture*. New York, NY: Taylor & Francis, 2008, pp. 168–187.
Treib, M. *Noguchi in Paris: The UNESCO Garden*. San Francisco, CA: William Stout Publishers, 2003.
Waldheim, C. "Landscape Urbanism: A Genealogy." *PRAXIS: A Journal of Writing + Building*, 4(2002), pp. 10–17.
Wall, A. "Programming the Urban Surface." In: J. Corner (Ed.). *Recovering Landscape: Essays in Contemporary Landscape Architecture*. New York, NY: Princeton Architectural Press, 1999, pp. 233–249.

6 Burle Marx
The Individual Language of Plenitude

Ana Rita Sá Carneiro

The greatness of an artist lies in the individuality of the language that they use to express themselves. It reveals their sensitivity in how they perceive the world through their poetic, musical, and other forms of representation. This essay explores the visual representation and graphic styles used by Brazilian landscape architect Roberto Burle Marx. The art critic Clarival do Prado Valladares states, "by observing and noting, exhaustively, overlapping plant structures, just as they pile up in Nature, Roberto Burle Marx was able to construct an individual language of plenitude in his drawings."[1] The title of this essay is inspired by this quote.

From a very young age, Burle Marx's restless and impetuous spirit seemed to embrace the challenges of life by exposing his willingness to change the paradigm and to let this process be governed by the forces of Nature. This willingness paved the way to his landscape creations and revealed his passion for garden design. This curiosity fostered a modern language—that is, the language of plants. Burle Marx shows that the plant, a living organism, is the primordial element that bursts forth at the start of his career; in his designs of gardens as concrete landscapes; and that he follows a process of decoding or geometrizing 'Nature' in his search for the "interior of matter," the DNA of vegetation.[2]

His language of landscape is represented through perspectives and in India ink, with the drawings of the first public gardens in the city of Recife, in the Northeast of Brazil in 1935. Vegetation as an artistic composition appears as a plastic element revealing details of the foliage of native plants, primarily as they are in physical reality so that specific species are identifiable as drawn. Over time, this representation acquires a visual appeal, in geometric shapes and colours, like abstract paintings. His drawings respond to a process of immersion in Nature, as if the shapes of the plants proposed for the garden were synthesized and decoded, adding to the landscape different pieces of information. Painting, botany, and music are all expressions of his artistry. In interviews throughout his career as a designer of landscapes, Burle Marx stated that he learned from plant scientists that a botanical association could also be an aesthetic association.[3]

His creative process and his understanding of Nature can be explained by his frequent excursions into the interior of the country to discover Brazil's flora. He became familiar with the habitat, which would later enhance his skills in painting and in other arts. For this reason, the graphic representation of the thematic landscapes of the garden is also a reflection of the artist's experience of regional landscapes and his responding concern with responding social, economic, and urban needs of the population at large. What should not be forgotten is the emphasis that he placed on the need for aesthetics because, according to Burle Marx, the garden "is neither a luxury nor a waste of space, but an absolute necessity for human life, without which civilization itself would lose its ethical reason."[4] In other words, his concern was to make art for people out of Nature, and in this sense, the garden is united through plants. "The existence of art is pinned to a necessity of life."[5] He understood the garden as a necessity for the human spirit, while his desire to know and learn from Nature was his reason for living.

Representations of a Garden

The representation of the garden as a tangible landscape came from Burle Marx's first professional experience as the Director of Parks and Gardens in 1935, when he was 25 years old, in the city of Recife, Brazil. At that time, though having trained as a landscape designer, he was still without professional recognition. His duties were those of a master-gardener who taught his assistants to perform the appropriate planting design. For this reason, he felt the need to draw in perspective, thus reproducing the 'particularities' of the plant to facilitate identification and composition at the time of the planting. He believed that it was only after such training that the assistants could tend accordingly to the maintenance of the garden. Local newspapers published his instructional statements where he explained, in detail, the conception of planning a garden. Burle Marx set out the historical, botanical, educational, social, architectural, and artistic arguments through these newspaper articles that also included his detailed drawings. Therefore, it was via his drawings that he sought to make himself understood to a mass audience. Detailing his design ideas in this manner allowed him to share his mission in introducing knowledge of the garden as "Nature organized and subordinated to architectural laws."[6] This amounted to combining plants that relate to each other and meet both scientific and artistic criteria. Overall, he managed to communicate the principles of his conception. "From an artistic point of view, the garden must obey a basic idea, with logical perspectives and subordinated to a certain *form of grouping*."[7] His reference to 'form of grouping' indicates the relationship between the parts—the notion of the garden's integrity as a whole system, considered as a *unit*. With this statement, Burle Marx reveals the principle of form and composition as a priority, which translates the garden's structure into a comprehensive plan, complete with volumes and textures of the selected plants. His aesthetic sensibility also intuitively defined a wide pathway along which users could observe the landscape, for as long as they wished, and enjoy the plant types and the variation that lighting conditions would provide, for example, their projection into the ponds (Figures 6.1 and 6.2).

The presentation of the idea and the design of the garden were teachings that would serve to guide gardeners and technicians alike and educated society about the meaning of each selected plant and its relationship in the grouping—the idea of grouping awakening an appreciation for the garden and respect for Nature. These narratives, materialized in the drawings, impress viewers by their comprehensive grasp of botanical knowledge in which he indicates particular species and their artistic qualities. A walk through the garden allows a visitor to experience the aesthetics of the space and engages all of their senses. This experience fosters the knowledge of landscape architecture in Brazil.

Burle Marx, when referring to the Casa Forte Garden—*Praça de Casa Forte*—in the city of Recife, explains the choice and origin of each plant, whether its Amazonian, American, or exotic. He talks about its decorative quality and how particular plants fit into the overall composition of the garden and details the relationship of the plants to other species. The resulting planting scheme created shading, provided an educational purpose, and followed his aesthetic criteria (see Figures 6.1 and 6.2).

Undoubtedly, his drawings were a means of communication that presented the tropical flora in the city with the purpose of communicating the enjoyment of the modern garden with an educational, environmental, and artistic function.[8] What existed until then can be best characterized as a European garden of exotic plants that were suggested by international engineers and not local landscape architects. While this design represented the garden in its 'plenitude,' it also addressed time in order to guarantee the maintenance of the garden and to consolidate the identity of the regional landscape and that of the Brazilian and the American landscapes.

With the cactus garden, *Praça Euclides da Cunha*, the design has a coherent and defined form. The geometrical and expressive shape of the cactus communicates a strong message and associates cacti with the hot climate and the vivid landscape of the tropics. For this reason, this image of the cactus must have registered in the artist's mind while observing it within the Botanical Garden of Dahlem in Berlin in 1928. In addition, Burle Marx states that reading the book *Os Sertões* (*The Backlands*), by the Brazilian writer Euclides da Cunha, impressed him with its poetic narrative about the exuberant landscape of the Caatinga biome in the dry land of the *sertão* region of northeast Brazil. There, during his first excursion, Burle Marx found the cacti growing among rocks and blooming in flower. The shape of the cactus in the landscape of the Caatinga intrigued Burle Marx to such an extent that he decided to bring them into an urban garden as if he were discovering the soul of Brazil:

6.1 and 6.2
Two drawings by Burle Marx for Praça da Casa Forte, 1935.
Source: Landscape Laboratory of the Federal University of Pernambuco (*Recife: Diario da Manhã*, 22(5), 1935, p. 1).

"It is urgent that we begin, right now, to sow, in our parks and gardens, the Brazilian soul."[9] The Cacti are constructive elements that are aesthetically pleasing surrounded by rocks of the regional landscape; at this particular garden, there was also the statue of an indigenous Brazilian, further reflecting the essence of Brazil (see Figures 6.3 and 6.4).

The Mystery in the Interior of Matter

The experience Burle Marx had in the *sertão* observing the Caatinga spurred him to distil the essence of the regional landscape for the moulded abstract form of the design of the roof garden of the Ministry of Education and Health (1938), a building that was a symbol of modern architecture in Rio de Janeiro.[10] Regarding the design of the roof garden, he explained in a statement in the newspaper

6.3 and 6.4
Two drawings by Burle Marx for Praça Euclides da Cunha (Cactario da Madalena), 1935.
Source: Landscape Laboratory of the Federal University of Pernambuco (*Recife: Diario da Tarde*, 14(3), 1935, p. 1).

Correio da Manhã that he idealized free forms of abstract painting based on what he absorbed from Nature, such as the drawing of leaves or a winding line of a riverbed as seen from the air:

> In the next two that I planned … I launched the experiment with free forms, merging plants and construction materials. They are gardens in which an abstract painting and a defined reality of space emerge united by floral forms related to both those elements, bringing to the viewer's mind the idea of Brazilian rivers, when seen from the window of an airplane.[11]

The winding line of a riverbed or the configuration of a broken boulder inspired the landscape designer as a 'plastic possibility' and an opportunity offered by Nature. There is speculation that the designs of the roof garden, which was represented in the vibrant colours which symbolized Burle Marx's style, and other projects such as the Army Square—*Praça do Ministério do Exército*—in Brasília, were specifically prepared for the exhibition at the Museum of Modern Art in New York in 1991, for which William H. Adams was the curator.[12] For various reasons, this style range allowed for the framing and the play of form and colours like a puzzle communicated visually with perfection[13] (see Figures 6.5 and 6.6).

It is the art critic, Pietro Bardi, who expresses that Burle Marx's intention to create free forms from what he finds in Nature, such as the apparent irrationality of the braided roots of a tree.[14] Bardi says that Burle Marx apprehends the forms of Nature—of the trunks, of the foliage, of the rocks—and starts to deconstruct them to extract their synthesis.[15] 'Un-configured,' geometrized, and colourful shapes are in the artistic composition of the landscaping and the painting (see Figures 6.5 and 6.6).

As a Brazilian art critic, Clarival do Prado Valladares believes that Burle Marx went from a phase in which his composition were based on the 'real figure' to the composition derived from the 'free form' of the planting skeleton as a plastic intention, thereby distinguishing him from his contemporaries of Picasso, Braque, and Matisse; this is reflected in the dialogue of the drawing between a painting and a garden.[16] Through his research, Valladares also believes that the rationale for Burle Marx's design is achieving the intimate structure of a chosen natural element, inspired by Braque's and Matisse's start from the plastic form of the isolated element (see Figures 6.7 and 6.8).

> I once took the trouble to confront him with numerous other artists from all ages. I managed to bring him closer to Braque and Matisse, in terms of motivation and formal research, but while these two great masters were motivated and started from the plasticity of the isolated element (Matisse), or from the composition still retained in the schema of a still life (Braque), the individuality of Roberto Burle Marx's drawing could be perceived in its origin, in its motivation, which is the intimate structure of the elements targeted *in natura*.[17]

The artist Roberto Burle Marx admits that he was the poet of his own life and wound his way through it with poems of lines, curves, and paints, each expression has its own *raison d'être*. Each phase has a meaning, which is confirmed by the opinion of the art critic Lélia Coelho Frota: "Burle Marx's graphic design, drawn largely in black-and-white, is as important as his colour work."[18] As different gestures, these representations today are reality; many of them are recognized as examples of Brazilian living heritage, starting with the mystery of the plant. For a landscape artist, the plant contains a historical process in of itself since it incorporates experiences in order to perfect colour, structure, and

6.5
Plan of roof garden, Ministry of Education and Health, 1938 (new drawing, 1990).
Source: All the rights reserved to Burle Marx Institute.

6.6
Plan of garden, Burton Tremaine Residence, Santa Bárbara, California, 1948 (new drawing, 1990).

© Copyright. Garden Design for Beach House for Mr. and Mrs. Burton Tremaine, project, Santa Barbara, California (Site plan). 1948. Gouache on board, 50 1/4 × 27 ¾ inch (127.6 × 70.5 cm). Gift of Mr. and Mrs. Burton Tremaine. Digital Image © The Museum of Modern Art/Licensed by SCALA/Art Resource, NY.

6.7
Plan of Ibirapuera Park, São Paulo, 1953 (new drawing, 1990).

© Copyright. and Oscar Niemeyer (Brazilian, 1907–2012). Ibirapuera Park, project, São Paulo, Brazil, Site plan, 1953. Gouache and graphite on board, 39 1/2 × 59 1/2 inch (100.3 × 151.1 cm). Gift of Roblee McCarthy, Jr. Fund and Lily Auchincloss Fund. Digital Image © The Museum of Modern Art/Licensed by SCALA/Art Resource, NY.

6.8
Plan of geometric garden in Ibirapuera Park (detail), São Paulo, 1953 (new drawing, 1990).
© Copyright. Ibirapuera Park, Quadricentennial Gardens, project, São Paulo, Brazil, Plan, detail five. 1953. Gouache on board, 43 × 52 1/8 inch (109.2 × 132.4 cm). Inter-American Fund. Digital Image © The Museum of Modern Art/Licensed by SCALA/Art Resource, NY.

shape and, therefore, is slotted into the "plane of aesthetic beings."[19] Looking for a form in Nature as a plastic intention of landscaping, Burle Marx builds a splendid, incomparable, indisputable, unique work, an international language of gardens.

Acknowledgements

I am grateful to Dr. Nadia Amoroso and Dr. Martin Holland for their invitation to contribute this chapter. I would also like to acknowledge the help of Centro Cultural Roberto Burle Marx, Instituto Burle Marx, José Tabacow, Julia Rey Pérez, Roderick S. Kay, and Wilson Feitosa. I would also like to acknowledge Sonia Berjman, who was a Senior Fellow and member of the Board on Landscape Studies of the Dumbarton Oaks Library and Researcher of the University of Buenos Aires: of the Faculty of Philosophy and Letters' Institute of Argentine Art History, of the Faculty of Architecture, Design, and Urbanism's Institute of American Art.

Notes

1. Valladares, C. "Roberto Burle Marx: Pintura em forma de jardim." In: P. Queiroz, L.V.P. Queiroz, and L. Boff (Eds.). *Roberto Burle Marx: Homenagem à natureza*. Rio de Janeiro: Ed. Vozes, 1979, p. 68.
2. "The Interior of Matter," collection of drawings and poems by Roberto Burle Marx and Joaquim Cardozo organized by Cristina Jucá and Gastão de Holanda in 1975.
3. Oliveira, A.R. "Bourlemarx ou Burle Marx?" Arquitextos. Web. 2 June 2001. www.vitruvius.com.br/arquitextos/arq013/arq 013-01.asp.
4. Affirmation made at a conference in 1954 called "Concepts of Composition in Landscaping." Burle Marx, R. *Arte e Paisagem: Conferências escolhidas*. São Paulo, Brazil: Nobel, 1987, p. 12.
5. Burle Marx, R. "Burle Marx fala de arte e paisagem." *Rio de Janeiro: Correio da Manhã*, 1935, 22(05).

6 Ibid.
7 Ibid.
8 Ibid.
9 Burle Marx, R. "Jardins e Parques do Recife." *Recife: Diario da Tarde*, 1935, 14(03).
10 Today the building is called Gustavo Capanema Palace.
11 Burle Marx, R. "Artes plásticas: Burle Marx fala de arte e paisagem". *Rio de Janeiro: Correio da Manhã*, 1954, 23(12): 14.
12 Adams, W. "Roberto Burle Marx: The Unnatural Art of the Garden." The Museum of Modern Art, New York, May 23–August 13, 1991, p. 7.
13 According to Vaccarino (2000, p. 47), the drawings of Burle Marx dialogue with the surrealist paintings of Jean Arp and Joan Miró.
14 Bardi, P. *The Tropical Gardens of Burle Marx*. Amsterdam, The Netherlands, Rio de Janeiro, Brazil: Colibris Editora Ltda, 1964, p. 16.
15 Sá Carneiro, A.R. "Roberto Burle Marx (1909–1994): Defining Modernism in Latin American Landscape Architecture." *Studies in the History of Gardens & Designed Landscapes*, 2019, 39(3): 255–270.
16 Valladares, *op. cit.*
17 Valladares, *op. cit.*, p. 68.
18 Frota, L.C. "A Painter and Visual Artist in the Brazilian Modernist Movement." In: R. Vaccarino (Ed.). *Roberto Burle Marx: Landscapes Reflected*. New York, NY: Princeton Architectural Press, 2000, p. 31.
19 Tabacow, J. *Roberto Burle Marx: Arte &Paisagem*. São Paulo, Brazil: Studio Nobel, 2004, p. 85.

References

Adams, W. *Roberto Burle Marx: The Unnatural Art of the Garden*. New York, NY: The Museum of Modern Art, 1991.
Bardi, P. *The Tropical Gardens of Burle Marx*. Amsterdam, The Netherlands and Rio de Janeiro, Brazil: Colibris Editora Ltda, 1964.
Burle Marx, R. *Arte e Paisagem: Conferências escolhidas*. São Paulo, Brazil: Nobel, 1987.
Burle Marx, R. "Artes plásticas: Burle Marx fala de arte e paisagem." *Rio de Janeiro: Correio da Manhã*, 23(12) (1954): 14.
Burle Marx, R. "Burle Marx fala de arte e paisagem." *Rio de Janeiro: Correio da Manhã*, 22(5) (1935): 14.
Burle Marx, R. "Jardins e Parques do Recife." *Recife: Diario da Tarde*, 14(3) (1935): 1.
Burle Marx, R. "O jardim da Casa Forte." *Recife: Diario da Manhã*, 22(5) (1935): 1.
Doherty, G. *Roberto Burle Marx Lectures: Landscape as Art and Urbanism*. Zurich, Switzerland: Lars Muller Publishers, 2020.
Eliovson, S. *The Gardens of Roberto Burle Marx*. New York, NY: Harry N. Abrams, Inc/Sagapress, Inc., 1991.
Frota, L.C. "A Painter and Visual Artist in the Brazilian Modernist Movement." In: R. Vaccarino (Ed.). *Roberto Burle Marx: Landscapes Reflected*. New York, NY: Princeton Architectural Press, 2000, pp. 25–32.
Frota, L.C. *Burle Marx: Landscape Design in Brazil*. São Paulo, Brazil: Câmara Brasileira do Livro (Brasiliana de Frankfurt), 1994.
Oliveira, A.R. "Bourlemarx ou Burle Marx?" In: Arquitextos. Web. 2 June 2001. www.vitruvius.com.br/arquitextos/arq013/arq 013-01.asp.
Pérez, J. "Burle Marx y su intervención en el paisaje cultural de Copacabana Documentación, análisis y protección de un patrimonio contemporáneo." PhD diss. (Universidad de Sevilla, 2012).
Sá Carneiro, A.R. "Roberto Burle Marx (1909–94): defining modernism in Latin American landscape architecture." In: S. Berjman and A. Tchikine (Eds.). *Studies in the History of Gardens & Designed Landscapes*, 39(3) (2019), pp. 255–270.
Tabacow, J. *Roberto Burle Marx: Arte &Paisagem*. São Paulo, Brazil: Studio Nobel, 2004.
Valladares, C. "Roberto Burle Marx: Pintura em forma de jardim." In: P. Queiroz, L.V.P. Queiroz, and L. Boff (Eds.). *Roberto Burle Marx: Homenagem à natureza*. Rio de Janeiro, Brazil: Ed. Vozes, 1979, pp. 67–70.

7 J.B. Jackson
Representing Everyday Landscapes

Jeffrey D. Blankenship

John Brinckerhoff (J.B.) Jackson (1909–1996) was not a landscape architect, nor was he a geographer or a historian; he was not a design professional or an academic in the traditional senses of these terms. Nevertheless, J.B. Jackson's work has influenced how landscape architects (and many other disciplines) think about and represent landscapes as lived human environments. With a bachelor's degree in history and literature from Harvard University (1932) and a fascinating early life of travel and adventure, it is a testament to Jackson's perambulating intellect that he was ultimately able to move, uncredentialed, among so many disciplines with an influence that is still felt 25 years after his death. In the small magazine that J.B. Jackson founded in 1951, *Landscape*, his influence was first and most prolifically felt by drawing attention to the *everyday* landscapes that humans have shaped over time. These everyday landscapes were not conjured in the imaginations of designers and planners; instead, they were the often-informal product of cultural, social, and economic processes and clever adaptations to the natural world. Jackson's curiosity was drawn to farms and feedlots, gas stations and the highway strip, trailer parks, bus stations, and cities' grittier quarters: places of little concern to mid-twentieth-century landscape architects navigating an expanding professional role in the post-war boom of urban and suburban development. Seeing the cultural landscape in all of its messy vitality required a particular open-minded curiosity. Critics of so-called visual blight were particularly inept at looking beyond their scorning critique to see the meaning embedded in the most prosaic of places, but Jackson insisted that his readers, and eventually students, learn to stop and think about what they were seeing before passing judgement.

Throughout Jackson's editorial tenure at *Landscape* magazine (1951–1968), many landscape architects were ultimately drawn to his message to take the ordinary and the unplanned seriously as a prerequisite to intervening in the places and lives of the "average citizen."[1] J.B. Jackson taught landscape architects to broaden their frame of reference beyond the narrow lens of professional practice and to consider the subject of landscape from the standpoints of history, geography, and culture. While *Landscape Architecture Quarterly* during the 1950s was strictly the professional magazine of the American Society of Landscape Architects (ASLA), Jackson's *Landscape* magazine effectively served as the de facto "academic" journal for the broader transdisciplinary field of landscape studies which Jackson had founded with his publication. J.B. Jackson's provocative essays—appearing in *Landscape* magazine and later in several edited collections—are widely acknowledged for how they taught his readers to see the everyday landscape. Perhaps of equal importance to landscape architects and other design professionals were the accompanying illustrations, photography, and overall graphic design of *Landscape* magazine, and eventually the images included in his edited collections and lecture slides.

Personal Reconnaissance

Jackson began drawing as a young man and had some formal training after Harvard when he briefly studied architecture at MIT, followed by some time in a commercial drawing school in Vienna.[2] Neither program satisfied his restless spirit for long, and he soon set off on a motorcycle tour of pre-war

DOI: 10.4324/9781003183402-7

7.1
Untitled drawing on brown paper: telephone poles, road, and railroad in the American Southwest, 1947, by J.B. Jackson. Reprinted with permission from the Collection of Helen Lefkowitz Horowitz.

Europe, but drawing—and occasionally watercolour—would be a lifelong habit. During World War II, Jackson served as an army intelligence officer, and he used simple drawings as a tool of reconnaissance to aid in planning troop movements through the landscapes of North Africa and Europe, establishing an efficient pragmatism to his sketches. Many of Jackson's travels and intellectual explorations would be recorded in notebooks and sketches for much of his life, as he conducted a sort of personal survey of the human landscape. Most of these surviving early drawings are in private collections and have only recently been published, which has provided welcome insight into his visual explorations.[3] As evidenced by these drawings, his style was quick, informal, and impressionistic. He would often draw with simple tools on whatever paper was available without aspirations to fine art. Figure 7.1 serves as an example of these qualities. The untitled drawing from 1947 on plain brown paper is animated and aggressive, rendered in thick coloured lines (suggestive of conté crayon or pastel), depicting an open roadscape in the American Southwest with telephone poles, electrical lines, and railroad tracks. Notably, Jackson does not omit the signs of human intervention in an otherwise natural landscape, a prelude to much of his work in *Landscape* magazine which would draw specific attention to how technological modernity was as ubiquitous in the rural landscape as it was in the cities.

Illustrating Landscape

Landscape was a graphically bold and visually appealing magazine that was entirely self-published by J.B. Jackson. Although the magazine would eventually boast a long list of luminary contributions from key critical thinkers in landscape studies, the first couple of years were almost entirely written by Jackson under his own name and several pseudonyms. Adding to this impressive literary feat, he also designed and often illustrated the magazine for the 18 years he served as editor. Cover art, paper choice, typography, photography, and the illustrations that would accompany the wide variety of articles—uninterrupted by advertisements—were curated with the eye of someone who cared about the magazine's aesthetic (Figure 7.2).

Readers with a less-nuanced understanding of landscape as a descriptive term might have been surprised by the subject matter and the wide variety of decidedly un-picturesque imagery of grain silos and agricultural cultivation patterns, historic villages and hydroelectric dams, public housing,

7.2
Composite of some of the common symbols Jackson used in Landscape to demarcate the variety of topics covered in the magazine, including geography, architecture, nature, agriculture, space and perception, and the highway landscape. Reprinted with permission of Paul F. Starrs and Peter Goin of the Black Rock Institute.

and shopping centres. The magazine was provocative for its inclusivity: the landscape was everywhere to be seen, interpreted, and appreciated, if only we would stop and consider the human meaning and lived experience of these places. In the magazine, Jackson often utilized his quick, expressive hand to illustrate this diversity of human landscapes. Figure 7.3 is a drawing that accompanied Jackson's essay from 1957, "The Stranger's Path," which described the unheralded parts of small cities devoted to travellers arriving by bus or train.[4] The drawing captures the cacophony of unregulated signage demanding the attention of the masses of anonymous working-class visitors as they navigate the unfamiliar quarters of strange cities. The overlapping linework, hastily rendered people, and sense of vibrating movement represented the antithesis of mid-century planning's idealized ambition for order, spatial expansiveness, and functional separation. Essays and drawings like these implied that despite the best efforts of planners and designers to engineer the modern urban environment according to the latest social and aesthetic theories, there was real life to be found in ordinary, uncelebrated places.

Occasionally, Jackson's drawings would grace the cover of the magazine. Figure 7.4 is a drawing from the front and back cover of the Winter 1957–1958 issue of *Landscape*, the same issue that features one of Jackson's most provocative essays, "The Abstract World of the Hot-Rodder." In the essay, Jackson suggests that:

> The discoveries of science and in particular the insights of artists and architects have made us familiar with changing concepts of space and matter and motion…But what is our reaction when

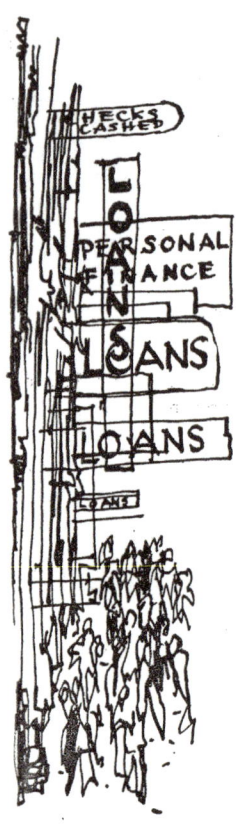

7.3
Drawing by J.B. Jackson from his essay "The Stranger's Path," 1957. Reprinted with permission of Paul F. Starrs and Peter Goin of the Black Rock Institute.

7.4
Front and back cover drawn by J.B. Jackson, Winter 1957–1958 issue of *Landscape*. Reprinted with permission of Paul F. Starrs and Peter Goin of the Black Rock Institute.

the man in the street tries in his own way to explore the same realm? We profess sympathy with the uncertainty, the inability to communicate of the contemporary artist; why do we express little or none for the hot-rodder and his colleagues?[5]

Although the drawing is not explicitly associated with that essay, there are qualities of abstraction and space and motion that seem to align with the article's discussion of the visceral experience of speed. Once again, the viewer is subjected to an image, on the cover in this case, that challenges any traditional notion of landscape, perhaps suggesting, along with the Hot-Rodder essay, that our perceptions and expectations of landscape are altered by modern concepts in science, art, and architecture.

The Autumn 1967 issue shown in Figure 7.5 is another cover illustrated by Jackson, showing a somewhat abstracted image of roads, fields, lots, and homes continuing uninterrupted into the infinite distance. Again, there is no explicit connection made to a specific article. Still, the Notes and Comments section at the beginning of the issue features the essay by Jackson, "To Pity the Plumage and Forget the Dying Bird," which describes the "neglected" and "mismanaged" parts of the American landscape that are common to all American small towns and their hinterlands.

> The street leading out of town passes several trailer courts, not very pretty ones, then it re-joins the highway; once more you are in the midst of well-cared for fields, winding valleys full of greenery; and off on the horizon, fifteen miles away, is another water tower, and presumably another town like the one behind you.[6]

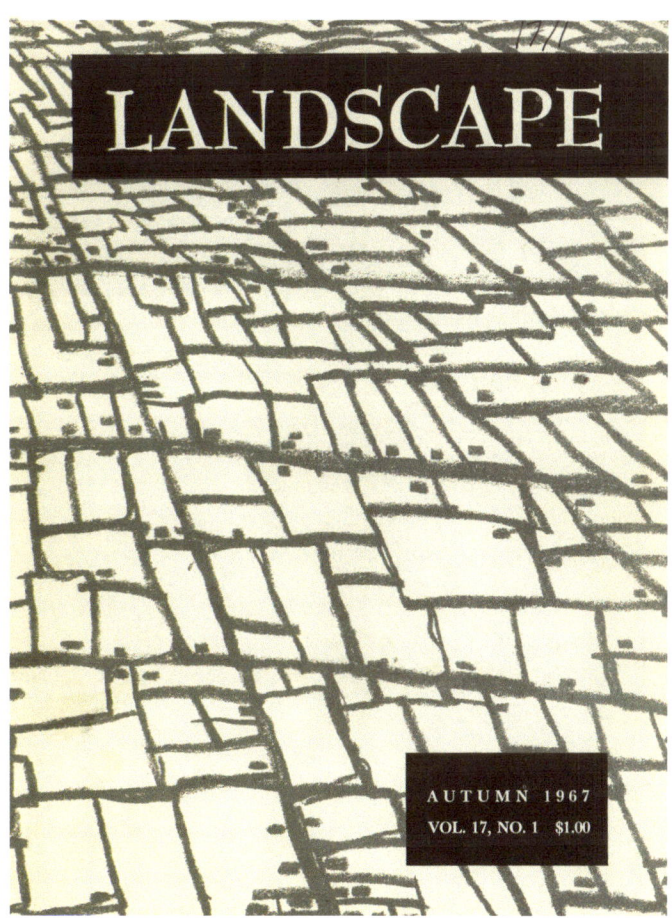

7.5
Front cover drawn by J.B. Jackson, Autumn 1967 issue of *Landscape*. Reprinted with permission of Paul F. Starrs and Peter Goin of the Black Rock Institute.

What comes through in the drawing is the sameness that Jackson often said he found fascinating about the American landscape; the sheer repetition of settlement patterns and vernacular architecture that were especially resistant to regional variation.

The View from Above

In the earliest issues of *Landscape*, J.B. Jackson called for articles and essays in the magazine to be accompanied by aerial photographs. Jackson's introductory essay in the first issue of *Landscape*, "The Need of Being Versed in Country Things," immediately worked to establish the usefulness of seeing the landscape from above by including an aerial photograph of the Rio Grande Valley near Alcalde, New Mexico, taken by the Soil Conservation Service. Jackson suggests, "It is from the air that the true relationship between the natural and the human landscape is first clearly revealed."[7] Figure 7.6 shows an early cover from the Autumn 1952 issue featuring an aerial photograph of contour terracing in the American Southwest, again by the Soil Conservation Service.[8] The landscape as viewed from above provided another form of landscape reconnaissance that complimented Jackson's drawings. Oblique aerial photography, in particular, aligned with Jackson's desire to highlight the cultural, environmental, and technological patterns that are uniquely visible from the air. This tool that would prove invaluable to landscape architects and planners in the coming decades and the development of Geographic Information Systems and remote sensing. Arguably, the book *Taking Measure Across the American Landscape* (1996) by the landscape architect James Corner and the aerial photographer Alex MacLean, which featured aerial photography, mapping, and essays, fits within

7.6
Front cover aerial photography by the Soil Conservation Service, Autumn 1952 issue of *Landscape*. Reprinted with permission of Paul F. Starrs and Peter Goin of the Black Rock Institute.

the lineage of Jackson's project to reveal the everyday modernity of the American landscape that is visible from above.

J.B. Jackson gave up the reins of *Landscape* magazine in 1968 and split much of his time between teaching courses on the American landscape at Harvard and Berkeley and producing edited collections of his essays. His lectures were immensely popular and often featured slides that Jackson took on his travels, often from the back of his motorcycle.[9] Many landscape architects were exposed to Jackson's work through these lectures and his books, which continued to argue for the importance of understanding the everyday (ordinary/common/vernacular/cultural) landscape. Jackson's essays could likely stand on their own without the illustrations and images discussed above, but capturing the unique visual qualities of the everyday landscape was important to J.B. Jackson personally and was an integral part of how he communicated his arguments. J.B. Jackson would continue to write and draw until his death in 1996.

Notes

1. J.B. Jackson often used shorthand phrases such as "the average citizen," "everyday people," or "the common man" to describe the inhabitants of the everyday landscape, as opposed to the elite decision-makers and planners that were shaping much of the modern landscape.
2. General biographical details of Jackson's life are drawn from Wilson, C. "A Life on the Stranger's Path," and Adams, F.D. "J.B. Jackson: Drawn to Intelligence." In: J. Mendelsohn and C. Wilson (Eds.). *Drawn to Landscape: The Pioneering Work of J.B. Jackson*. Staunton, VA: George F. Thompson Publishing, 2015; and Horowitz, H.L. "J.B. Jackson and the Discovery of the American Landscape." In: H. Horowitz (Ed.). *Landscape in Sight: Looking at America*. New Haven, CT: Yale University Press, 1997. For the most recent and detailed treatment of Jackson's life, see Horowitz, H.L. *Traces of J.B. Jackson: The Man Who Taught us to See Everyday America*. Charlottesville: University of Virginia Press, 2020.
3. See Mendelsohn, J. and Wilson, C. (Eds.). *Drawn to Landscape: The Pioneering Work of J.B. Jackson*. Staunton, VA: George F. Thompson Publishing, 2015.
4. Jackson, J.B. "The Stranger's Path." *Landscape: Magazine of Human Geography*, 1957, 1(Autumn): 13.
5. Jackson, J.B. "The Abstract World of the Hot-Rodder." *Landscape: Magazine of Human Geography*, 1957–1958, 2(Autumn): 13.
6. Jackson, J.B. "Notes and Comments: To Pity the Plumage and Forget the Dying Bird." *Landscape*, 1967, 1(Autumn): 1.
7. Jackson, J.B. "The Need of Being Versed in Country Things." *Landscape: Human Geography of the Southwest*, 1951, 1(Winter): 25.
8. The sources for many of the early images in *Landscape* were often from the collections of the Soil Conservation Service, the Farm Security Administration or the Standard Oil Company, which all undertook photo-documentary projects during the depression.
9. Notably, many of Jackson's slides have been recently digitized and are now searchable through the Center for Southwest Research's online archives. https://elibrary.unm.edu/cswr/.

References

Adams, F.D. "J.B. Jackson: Drawn to Intelligence." In: J. Mendelsohn and C. Wilson (Eds.). *Drawn to Landscape: The Pioneering Work of J.B. Jackson*. Staunton, VA: George F. Thompson Publishing, 2015, pp. 33–40.

Corner, J. and MacLean, A. *Taking Measure Across the American Landscape*. New Haven, CT: Yale University Press, 1996.

Horowitz, H.L. "J.B. Jackson and the Discovery of the American Landscape." In: H. Horowitz (Ed.). *Landscape in Sight: Looking at America*. New Haven, CT: Yale University Press, 1997, pp. ix–xxxi.

Horowitz, H.L. *Traces of J.B. Jackson: The Man Who Taught Us to See Everyday America*. Charlottesville: University of Virginia Press, 2020.

Jackson, J.B. "Notes and Comments: To Pity the Plumage and Forget the Dying Bird." *Landscape*, 1967, 1(Autumn): 1.

Jackson, J.B. "The Abstract World of the Hot-Rodder." *Landscape: Magazine of Human Geography*, 1957–1958, 2(Autumn): 13.

Jackson, J.B. "The Need of Being Versed in Country Things." *Landscape: Human Geography of the Southwest*, 1951, 1(Winter): 25.

Jackson, J.B. "The Stranger's Path." *Landscape: Magazine of Human Geography*, 1957, 1(Autumn): 13.
Mendelsohn, J. and Wilson, C. (eds.). *Drawn to Landscape: The Pioneering Work of J.B. Jackson*. Staunton, VA: George F. Thompson Publishing, 2015.
Wilson, C. "A Life on the Stranger's Path." In: J. Mendelsohn and C. Wilson (Eds.). *Drawn to Landscape: The Pioneering Work of J.B. Jackson*. Staunton, VA: George F. Thompson Publishing, 2015, pp. 19–30.

8 The EDSA Style
"A Legacy of Graphic Communication"

Kona A. Gray

The Origin

At EDSA, land planning, landscape architecture, and urban design have been expressed with sketches and hand drawings from the beginning of the practice. The firm launched in September 1960 when the profession of landscape architecture was emerging in newfound significance. *Design with Nature*, written by Ian McHarg and released in 1969, set an influential precedent for designing more than just parks and residential communities. It was a time of expressive design that only hand drawing could illustrate, propelling landscape architects such as Hideo Sasaki, Lawrence Halprin, and Garrett Eckbo to notoriety. The simple black and white plan, section, and perspective drawings brought the imagined landscapes to life. Techniques to provide depth without colour, including shade and shadow, were essential. Many of the earliest EDSA drawings expressed the nature of this sensibility, evoking simplicity, and readability for the viewer. As a form of communication, drawing is carefully utilized to take an idea from vision to reality. Analysis Diagrams, Concept Sketches, and Construction Documents were all developed with a limited palette to avoid mysteries in the drawings. These drawings are observations illustrated for ongoing contemplation, design exploration, and implementation.

The early drawings demonstrate how graphic representation has evolved, but ultimately, the purpose of the drawing remains the same.

> Throughout the firm's history, a high level of importance has been placed on the ability to express ideas through the simple medium of pen on paper. Proven time and time again, clients relate to the personal quality of presentation drawings produced by hand and respond to the unique character that is created in an original illustration.

These words from the *EDSA Graphics Book* put it all into perspective.[1] There is a practice of thoughtful observation that provides insight as you study a place. In the 2014 edition of *Design Matters*, an annual design magazine published by EDSA, Greg Kaueper explains:

> I stood within monumental spaces that make you feel tiny and insignificant. I walked down narrow streets and through small piazzas that were proportionally very much human-scale—all the while trying to determine the qualities that made these places sustainable. For my resources, I had places to walk, to eat, to sit, to sketch, to live. I had people to watch, talk with, and learn from. I took classes and lessons in language, cooking, and wine in order to gain an appreciation and understanding for the important elements that represent Italian culture.[2]

He was describing a trip to Italy where the people and the landscape are rich subjects for discovery. The EDSA Style reflects the intuitive skill of observation through the sketch to document what the eye has captured. In the 2013 video, *Laurie Olin on Design: Drawing is a Powerful Tool*, from the Cultural Landscape Foundation, he describes how

> one of the most amazing ways to learn about the world is to wander around in it and actually do drawings…to draw it. Because drawing it makes you look at it and sit still… and to think about what is in front of you.[3]

As landscape architects and designers, we can see life through the medium of drawing. Drawing is a tool that can be utilized anywhere. We see the most meaningful discussions flourish during design charrettes because of a drawing. Olin describes this as the "power of the pencil."[4] Drawing together and especially drawing in front of Clients is inclusive as well as meaningful. Everyone has an opportunity to contribute to the conversation of exploring together through the drawings. The initial drawings should be loose, simple, and rough. They are not precious. The drawings are meant to evoke ideas and questions.

Documenting stories, providing direction, and visualizing the world with a sketch have contributed to the essence of why drawing matters. The intent of a drawing is to initiate a dialogue regarding a place or a subject. To ask questions like "Why is this place so special?" and "What emotions do you feel in the space?" Hand sketching allows you to explore ideas to communicate important aspects of life. The ability to draw is essential for non-verbal communication, and it contributes to social understanding. However, drawing well does not always equate to good communication or even good design. So, how important is drawing in the process of design? Does drawing matter? In my opinion, drawing really matters! To be clear, it is not about just learning how to draw but instead using graphic representation to stretch your creative energy. It is exhilarating to see how drawing has evolved through the influence of technology and the inspirations that curate the story.

The Evolution

The *EDSA Graphics Book*, published in 2010, brought attention to the work that we find natural. After 50 years of practice, this culmination revealed the beauty of the craft that has been dismissed by many in exchange for computer graphics. We find value in the process and evolution by utilizing every tool. Our ability to hand draw translates directly into the computer via a digital tablet. Analogue and digital have been fused together at EDSA, and the foundation of the hand sketch remains in place as a springboard. The book notes how "a high level of importance has been placed on the ability to express ideas through the simple medium of pen on paper." We know that it works because it has been "proven time and time again, clients relate to the personal quality of presentation drawings produced by hand and respond to the unique character that is created in an original illustration." We utilized the book to solidify the importance of hand drawing.

> Hand graphics are synonymous with EDSA. While the concept is simple, the methods and mediums vary. From the beginning, a simple pen on tracing paper method has been the basis for most all hand graphics at EDSA. Over the years, new technologies have been introduced and combined with our time-tested graphics, to produce a unique look and feel. At EDSA, digital tools regularly provide a framework from which a graphic is produced, and indeed often finish the graphic as well. What happens in between is the crafting of a story in graphic form, where character, detail, and visual interest are added by the artist's hand.

The methods are broken down by type, including 'Pen, Marker on Trace, Marker on Bond, Water Colour, Colour Pencil, Marker with Digital Effects, and Digital Combinations.' It was discovered in the 1960s that pen and ink are just simple and quick. Also, there is no eraser! We believe that committing to the sketch increases your confidence, allowing you to focus more on the subject and less on the technique. The Pentel Sign Pen is a favourite drawing tool because it will enable you to create different line weights through the angle of the pen. When you hold the pen at 90 degrees, the line tends to remain thin. If you change the angle to 120 degrees, the line becomes medium. Finally, changing the angle to 160 degrees creates a wide, broad line. A precise drawing is not possible with a wide line-weight, forcing you to be loose and creative. This method travels well, too. Most designers at EDSA carry a roll of trace, a scale for quick measurements, and several pens to draw on demand. "When an idea needs to be explained or expanded upon during a client meeting or workshop, pen and trace paper are easily accessible and understood in any language." The digital tablet allows this freedom too. But you need to get comfortable drawing on glass instead of paper.

The EDSA Style has experimented with markers on trace and bond over the decades. It is a commitment to expand upon a beautiful pen and ink drawing by adding colour. We utilize a palette of "go-to" marker colours to compliment the black and white drawing. "Marker on trace is one of the quickest and most efficient techniques for producing illustrations." You can even create subtle colours with markers by rendering them on the back of the trace. During the 1970s, marker on bond introduced a new method that would allow for more bold colours than markers on trace. "Once the line work is completed, the pen drawing can be scanned, scaled, and modified as needed, and then plotted onto heavyweight bond paper." Next, designers could add colour without the worry of ruining an original drawing. "Unlike trace, bond is not forgiving in colour choices, however it is easy to plot multiple copies so as to test colour before the final rendering." Before to Adobe Photoshop, we would make revisions by "splicing" a new graphic into a plan like a surgeon. It was challenging to tell that changes were made. This process was also utilized on several renderings from the same decade to reveal bold colours balanced by negative space left when the paper is not rendered.

Although pen and marker have always been the method most utilized, the watercolour and colour pencil became unique overlays that added more texture. "Watercolours provide a soft, loose, and distinctive character that is hard to match with any other medium." Similar to markers on bond, you could print a black and white drawing onto watercolour paper to preserve the original trace. Like watercolour, colour pencil increased its presence in our drawings between the 1980s and 2000s. First, as a layer to bring out highlights in the pen and marker drawings. "A white coloured pencil is especially useful for providing sharp, bright, contrasting highlights as a final layer to a marker illustration." Next, we started utilizing the technique as the primary rendering method. Unfortunately, "coloured pencil is not the most efficient technique for producing a drawing. The method proved to be time consuming to lay the colour down, and the pencils have a shorter lifespan compared to markers." We learned that every drawing tool has an effective purpose.

The arrival of digital rendering was a game changer after 2010. At first, we utilized the hybrid aspect of scanning finished renderings into the computer to enhance with Adobe Photoshop.

> After a marker rendering has been created and scanned into a digital form, the overall look can be manipulated using a variety of computer programs. Increasing vibrancy in colours and highlighting shadows are some simple digital effects that can give a marker rendering more depth.

This period also ushered in considerable debate. The analogue-versus-digital conversation gained quite a bit of momentum.

> Although there are many advanced computer programs that could be used to render the entire drawing from start to finish, we find when the base drawing is done by hand and digital effects are simply used to enhance, it provides the best of both worlds.

The Future

The digital combination is the way forward. Today, we continue to utilize the sketch as a spark that initiates the idea. Then, we employ simple massing models in Rhino to further explore the idea. The model becomes the base for sketch overlays on the tablet as an iterative tool. "Once the designer is comfortable with the vignette, the base can be plotted and sketched over to create a warmer feeling. Finally, any variety of colour rendering techniques can be applied as desired to achieve a final product." In the 2015 edition of *Design Matters*, Eric Propes describes how

> we're always looking for new and emerging technology and exploring ways to work more efficiently. However, the truth is that all the technological advances in the industry cannot replace or devalue the freedom of hand sketching. Instead, we enhance our hand graphics by using technology as a complement to the design process, when and where it is most appropriate. Design

visioning is built around our ability to express ideas. That is the essence of our profession. And one of the best ways to express design—is to illustrate it.[5]

The EDSA Style will continue to evolve. It is an exciting time to be a designer, and the future of design has infinite possibilities. Refer to Figure 8.1, which shows Edward Stone with his landscape architecture team, and to Figures 8.2–8.8, which illustrate the typical graphic styles from EDSA.

8.1
Edward Stone with his landscape architecture team. Ed Stone understood hand-to-eye coordination developed through the process of drawing allows us to gain a deeper understanding of the subject. The vision inspired EDSA to establish a unique drawing style.

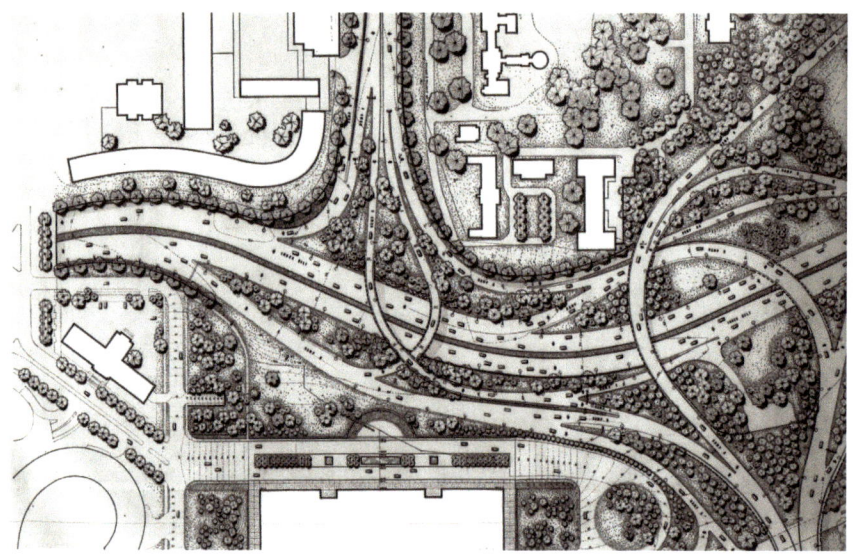

8.2
Renderings from the John F. Kennedy Performing Arts Center designed in early 1960s demonstrate the power of pen and ink. No matter if it is a detailed plan rendering or quick perspective, the idea is always clear.

The EDSA Style 63

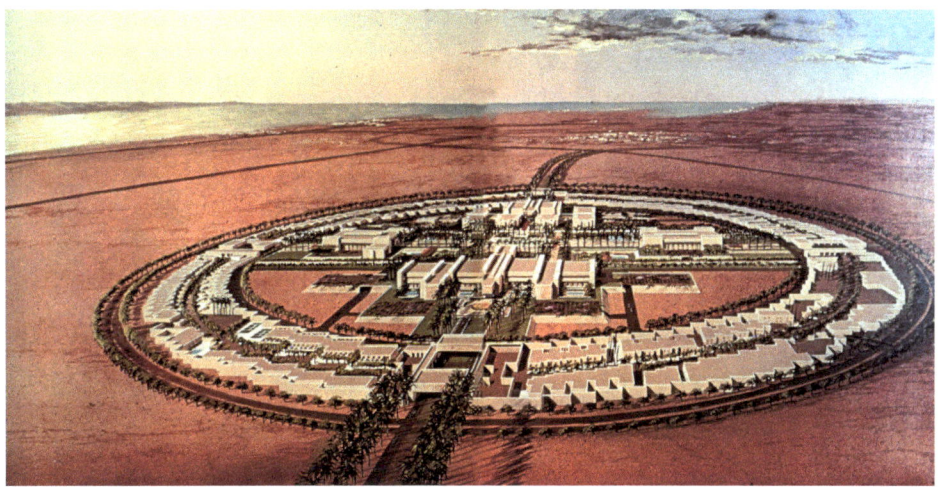

8.3
The Sunrise Saudi Arabia masterplan designed in the 1970s captures a rich sunset juxtaposed with the desert and Red Sea to evoke a very romantic expression.

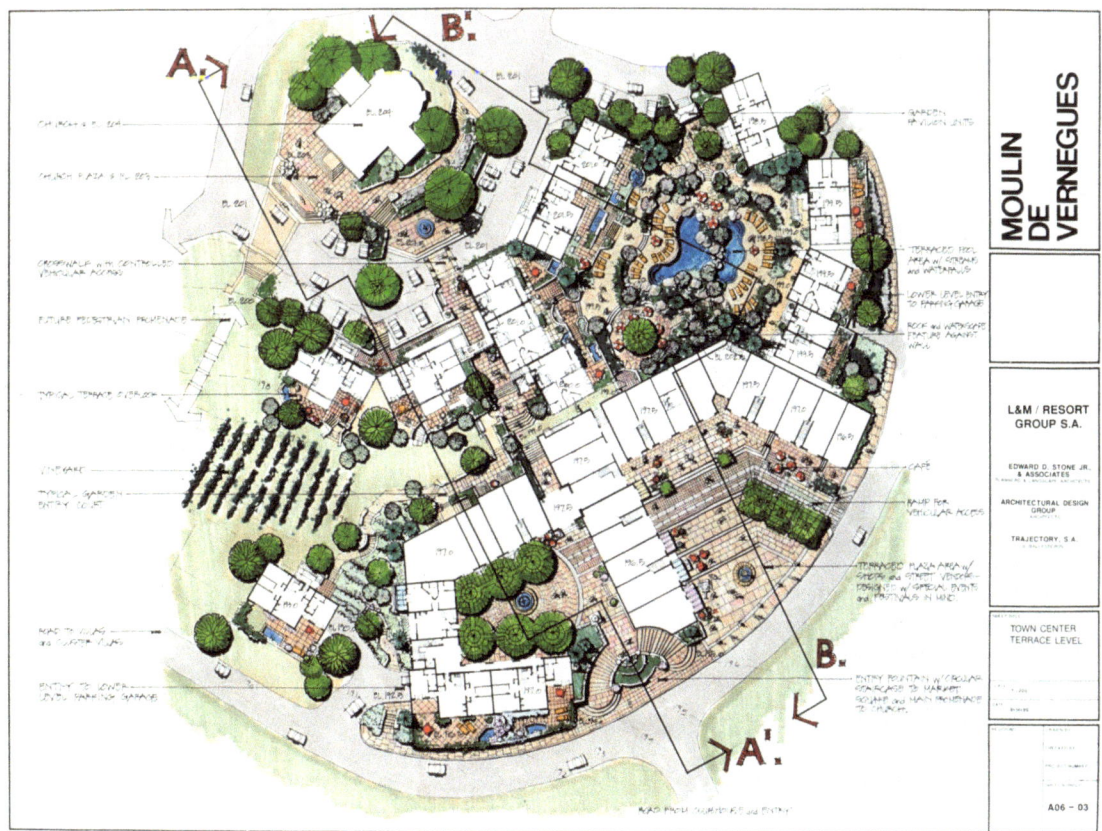

8.4
The "quick" marker technique on bond process was a metamorphosis which produced a similar effect as watercolour. The renderings from Pont Royal designed in the 1980s utilize this technique very effectively.

8.5
In the 1990s, projects like Blockbuster Park perfected the rapid marker "watercolour style" technique. This rendering style continued for the duration of the decade.

8.6
In the 2000s, watercolour became fully embraced as a softer method of artful communication. The Mohammed Bin Zayed City Master plan rendering is a good example of how the illustration provides softer tones.

The EDSA Style 65

8.7
Digital rendering opened the potential of the final drawing. It is like post-production in the film industry, as illustrated in the Gilgamesh Island perspective. The water and sky take on a new life with digital effects.

8.8
The Escuintla master plan rendering represents hand drawing over a model, layered with GIS, Aerial Photography and Adobe Illustrator and Photoshop to realistically capture the transformations. The digital combination is here to stay.

Notes

1. Quotes throughout the chapter come from the *EDSA Graphics Book*, unless otherwise noted: Smith, D.C. *EDSA Graphics Book*. Fort Lauderdale, FL: EDSA, 2010.
2. Kaueper, G. "The Practice of Observation." *Design Matters Magazine*, 2014: 17.
3. Laurie Olin, interviewed by C. A. Birnbaum, in "Laurie Olin on Design: Drawing as a Powerful Tool." YouTube video, 3:08. The Cultural Landscape Foundation, November 12, 2013, https://youtu.be/U5el6zo8kys.
4. Ibid.
5. Propes, E., Bulemore, D., and Cissel, D. "Digitally Designed or Not?" *Design Matters Magazine*, 2015: 55.

References

Kaueper, G. "The Practice of Observation." *Design Matters Magazine*, 2014, 3: 14–19.

Olin, L., interviewed by C. A. Birnbaum. "Laurie Olin on Design: Drawing as a Powerful Tool." YouTube Video, 3:08. The Cultural Landscape Foundation, November 12, 2013. https://youtu.be/U5el6zo8kys.

Propes, E., Bulemore, D., and Cissel, D. "Digitally Designed or Not?" *Design Matters Magazine*, 2015, 4: 54–61.

Smith, D.C., *EDSA Graphics Book*. Fort Lauderdale, FL: EDSA, 2010.

9 Boomerangs, Zig-Zags, and Orbits

Drawing the California Garden Garrett Eckbo and Thomas Church

Chip Sullivan

Biomorphic Heroes

I am thrilled to have this opportunity to discuss the drawings of two of my heroes, Tommy (Thomas) Church and Garrett Eckbo. I have always admired their visionary designs and drawings from the 1940s to 1950s. The summer before I entered the landscape architecture program at the University of Florida, my mother bought me a used copy of Eckbo's *Landscapes for Living* at a garage sale for $2.50. I immediately fell in love with the strange biomorphic garden illustrations and their complex cubist forms. However, as an undergraduate I could not understand a word of the text, which further led me to focus on the illustrations and even attempt in my studio courses to design in that style. In retrospect, I see that, throughout my career, these drawings have been a source of creative inspiration for me.

Westward Ho

Thomas Church, ten years senior to Eckbo, was born in Boston, Massachusetts. Not long afterwards, he moved with his family to the lush Ojai Valley in southern California and eventually settled in Northern California in Berkeley. Garrett Eckbo was also born "back east," in Cooperstown, New York. At an early age, he and his family moved to Alameda, California. Both men received their undergraduate degrees in landscape architecture from the University of California, Berkeley, and went on to obtain masters' degrees in landscape architecture from Harvard University.

Beaux-Arts Beginnings

Both Church and Eckbo were trained in the beaux-arts tradition at UC Berkeley. Beaux-arts pedagogy emphasized the process of production from quick conceptual sketches to highly finished presentation drawings. The typical curriculum would include instruction on figure drawing, sculpture, classical ordering systems, proportion, and presentation rendering. The mastery of freehand sketching and watercolour techniques were core learning outcomes, as was demonstrating competency in programming, site analysis, and spatial composition. Additionally, the quick tempo of the *esquisse*, in which students were given one week to develop a design from concept to presentation drawing, was excellent preparation for the rigorous pace of the professional office.

Thomas Church: Elegance and Artistry

Refer to Figure 9.1, Preliminary Plan for Mr. & Mrs. Charles Shuey, Claremont, CA, 1937. In this early presentation plan by Church, we can clearly see the influence of his beaux-arts training in its

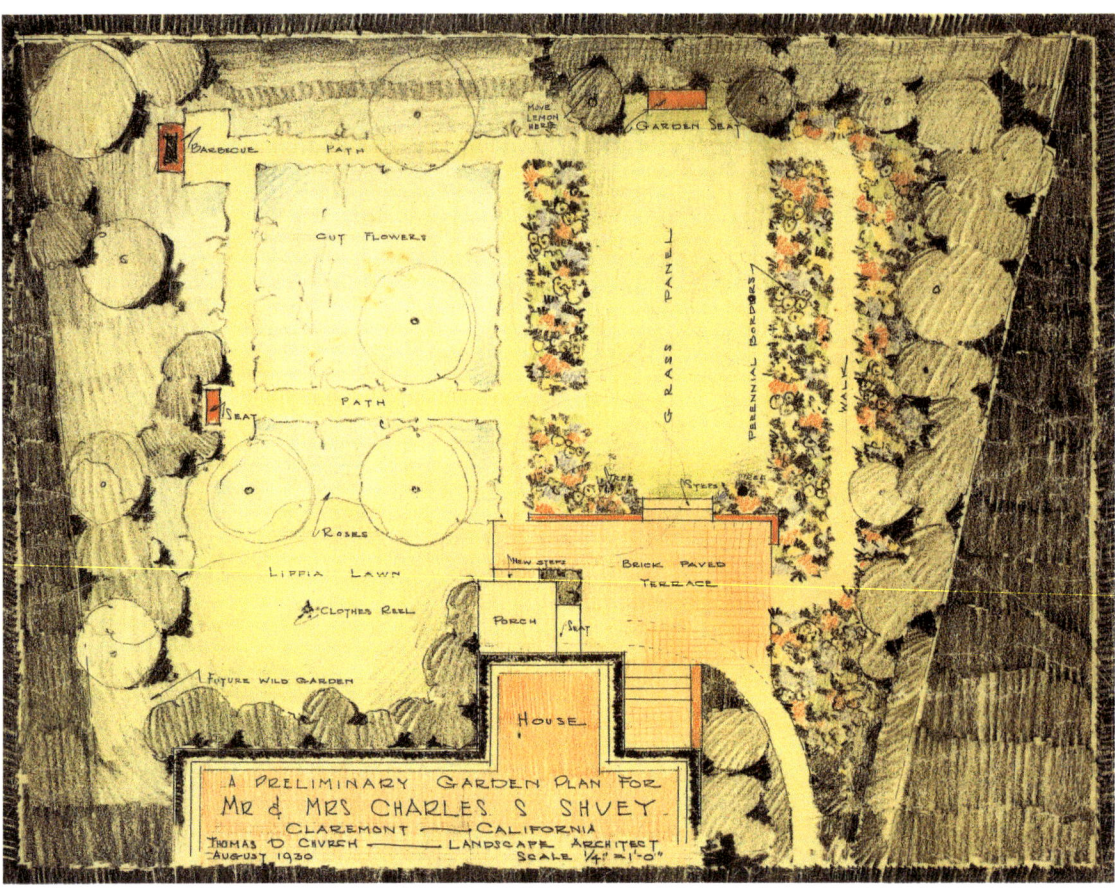

9.1
Preliminary Plan for Mr. & Mrs. Charles Shuey, Claremont, CA, 1937. By Thomas Church. Environmental Design Archives, University of California Berkeley.

spatial organization and geometry. Church's preferred drawing tools were an Eagle 314 Draughting Pencil and a grease pencil. The 314 pencil produced soft, flowing, satin-like lines as well as bold, dark expressive tones. The grease pencil is made of hardened black wax and is excellent for creating dark accents and shadows. The plan contains subtle variations of line-weight making the drawing come alive. Church used a chisel-pointed pencil to indicate the mass of the tree canopies with broad, soft, parallel lines. He used a grease pencil for the deep shadows and parallel hatches on the borders. He created light washes of colour and texture with Prismacolor pencils and accentuated the flowers. His beautiful hand-lettering complemented this masterful ensemble.

Over the years, the composition of the 314 pencil changed from lead to graphite. It no longer produces the same type of soft, inky lines as the original. Ironically, vintage Eagle 314 Draughting Pencils can go for $40–$50 a dozen on eBay today.

A New Vision

Refer to Figure 9.2, Preliminary Plan for the Martin Garden, Aptos, CA, 1947. After a trip to Europe in 1937 to study the work of architects Le Corbusier and Alvar Aalto, Church was inspired to experiment with new garden forms. Here, we can see this influence through his distinctive use of biomorphic forms, expressed with a loose, almost spontaneous drawing style. The drawing abounds with energy as loopy lines form the planting beds, wavy lines outline the hedges, 'Pollock-like' gestures define the tree canopies, and quick stippling illustrates the sand area. One can almost visualize Church's dynamic design process through his drawing technique.

Refer to Figure 9.3, Zelinsky Garden, Atherton, CA, 1947. This perspective sketch presents a slightly elevated view, which seems to be inspired by the drawing style of Le Corbusier. Church

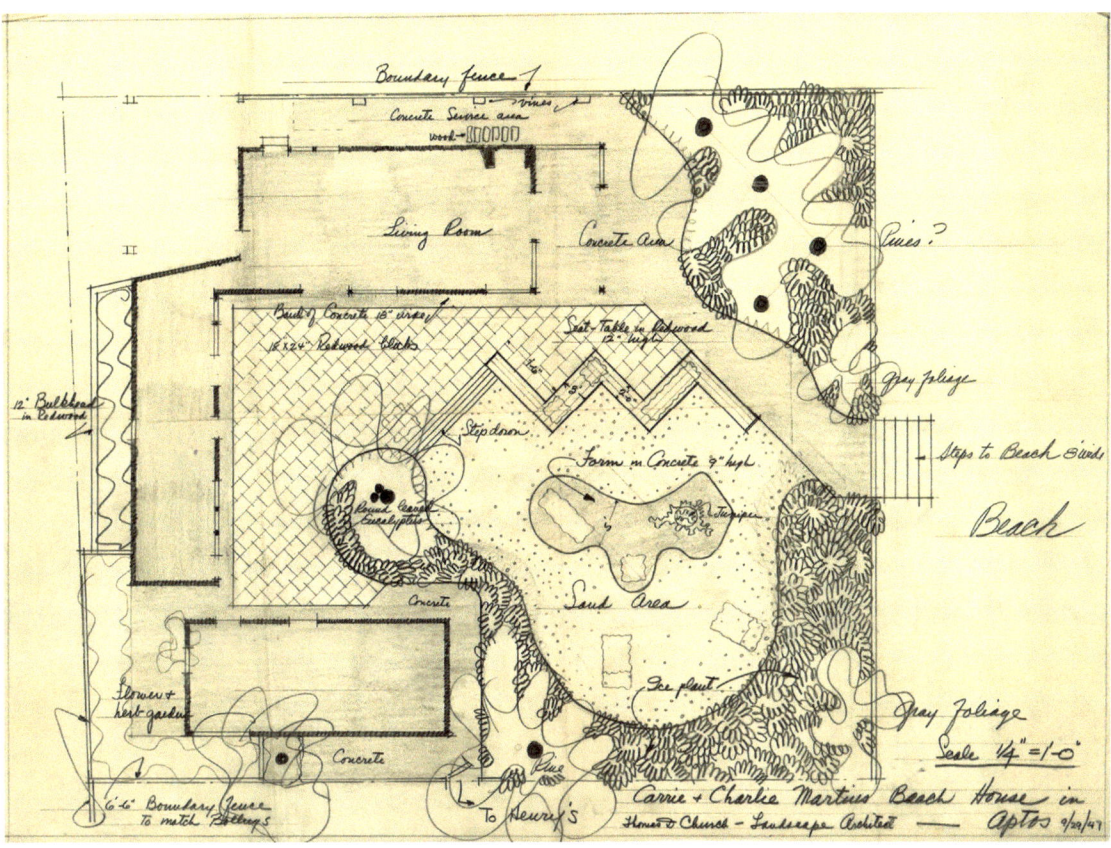

9.2
Preliminary Plan for the Martin Garden, Aptos, CA, 1947. By Thomas Church. Environmental Design Archives, University of California Berkeley.

9.3
Zelinsky Garden, Atherton, CA, 1947. By Thomas Church. Environmental Design Archives, University of California Berkeley.

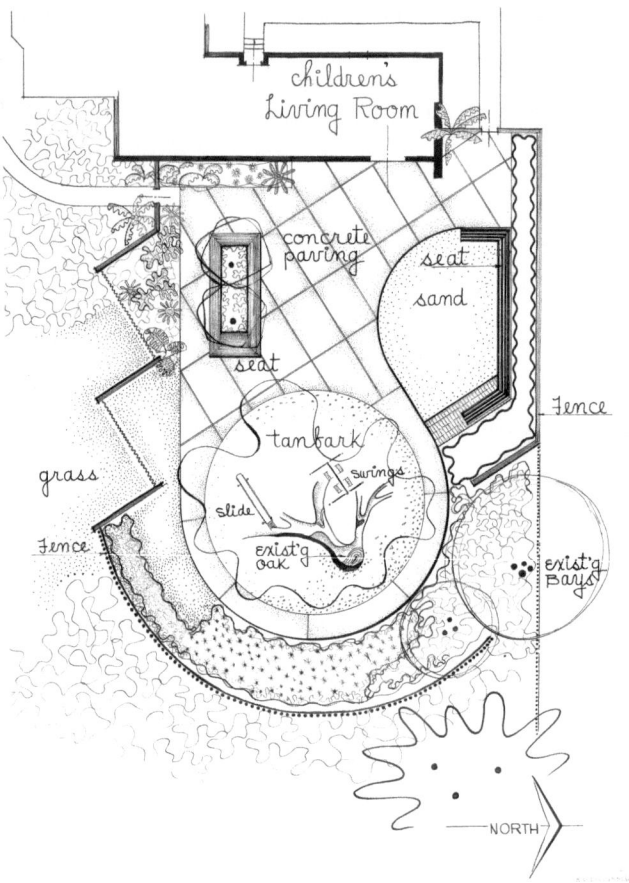

9.4
Hickingbotham Garden, Hillsborough, CA, 1948. By Thomas Church. Environmental Design Archives, University of California Berkeley.

abstracted the shapes and forms of plants with an economy of freehand lines and stipples for grass. With these few gestures, he created an expansive landscape space that captures the essence of the garden. Church primarily worked on Canary Yellow tracing paper, a very thin but sturdy, low-opacity paper. Its transparency makes it easy to trace drawings, produce many overlays, and most importantly, visualize the design process.

Refer to Figure 9.4, Hickingbotham Garden, Hillsborough, CA, 1948. This ink drawing is from Church's book, *Your Private World*, published in 1969. The variety of wavy lines and the plan view of the tree trunk and its canopy form an elegant visual composition. Intense stippling indicates the grass. The unique chevron shape of the north arrow and the loopy hand-lettering add an inviting modernist touch.

Garrett Eckbo: Escape from the '*Ancien Régime*'

Refer to Figure 9.5a, An Estate in the Manner of Louis XIV, 1934. This exquisite rendering was produced while Eckbo was an undergraduate in UC Berkeley's landscape architecture program. The site plan shows his early mastery of balance, proportion, and scale and exhibits a strong sense of spatial organization and subtle grading.

Beaux-arts illustrations like this were typically drawn on 140 lb. hot press watercolour paper. After the drawing was completed in pencil or permanent ink, the paper was soaked in water and then stretched onto a board and secured with paper tape. Once the paper dried, the colours were painstakingly applied in light washes. After each wash had dried, another layer of colour was applied to build up the tones. This was an arduous process but one that resulted in students developing their individual

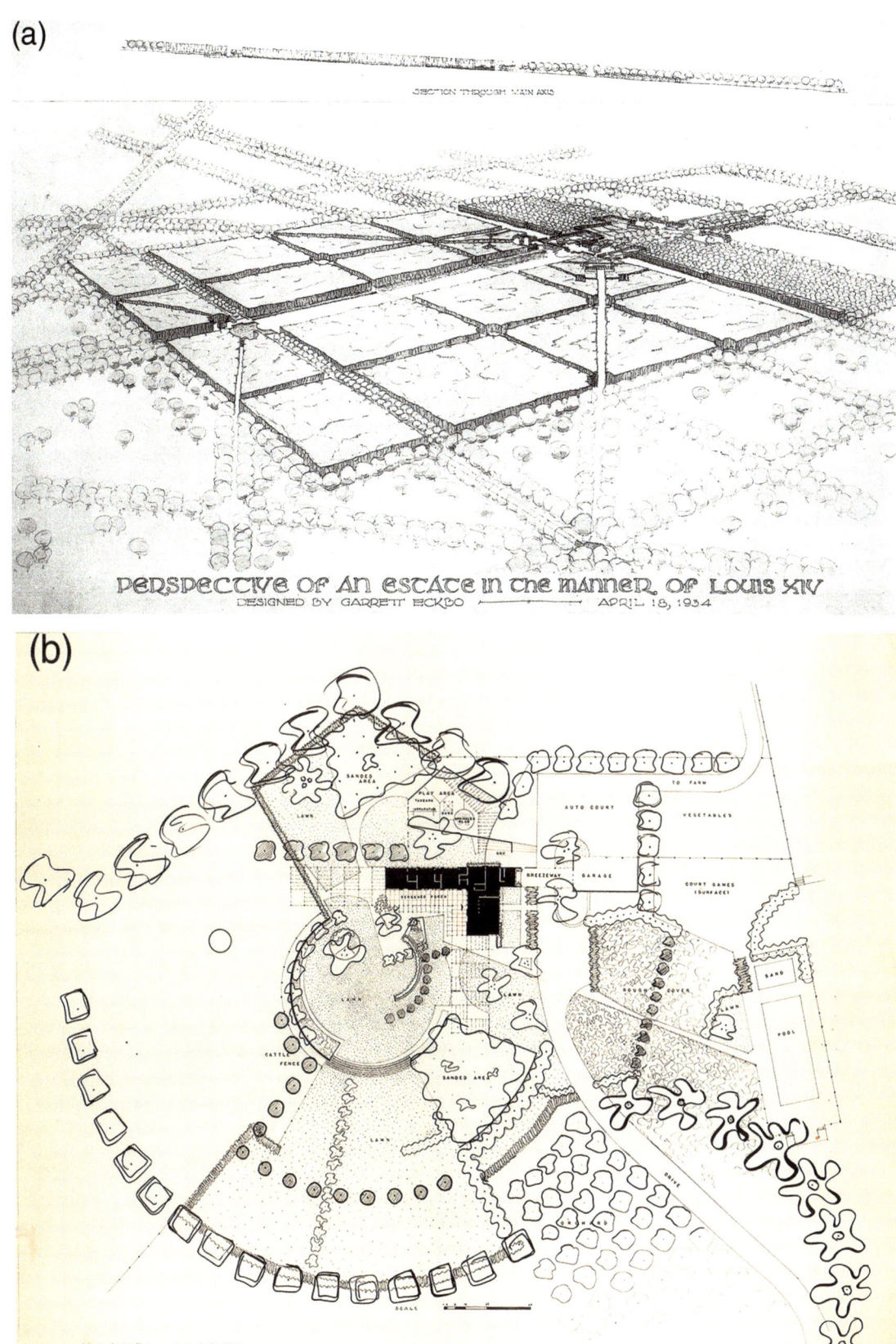

9.5
(a) An Estate in the Manner of Louis XIV, 1934. By Garrett Eckbo. Environmental Design Archives, University of California Berkeley. (b) Nickel Garden, Los Banos, CA, 1944. By Garrett Eckbo. Environmental Design Archives, University of California Berkeley.

painterly skills. In my many discussions with Garrett, he often emphasized his distaste for his formal beaux-arts training. He frequently mentioned how much he hated this method of rendering in graded washes, but I believe it was an essential foundation for his later innovations.

Jazz, Cubism, and Modernism: A Creative Explosion

When Eckbo entered the Harvard Graduate School of Design, the landscape architecture program still had remnants of the beaux-arts system, but the curriculum was quickly evolving. When Walter Gropius and Marcel Breuer joined the architecture faculty in 1937, they brought with them the pedagogy of the European Bauhaus. Eckbo immersed himself in the new theory of modern design and developed an interest in jazz and cubism as well. He and his fellow students came to reject the dictums of the *'ancien régime'* represented by the beaux-arts style.

Bombastic Boomerangs, Amoebas, and Zig-Zags

Refer to Figure 9.5b, Nickel Garden, Los Banos, CA, 1944. Before I moved to California in the late 1980s, I was fortunate to teach with Peter Walker as his apprentice at the Harvard Graduate School of Design. As I was wandering through the stacks of its amazing design library, I serendipitously came across a catalogue from the 1925 *Exposition Internationale des Arts Decoratifs Moderne* in Paris. As I leafed through it, I was surprised to see that the last person to check it out was Garrett Eckbo in the 1940s. He had renewed it many times. I could imagine how the innovative cubistic gardens depicted in the catalogue must have influenced him.

If you compare this drawing with the previous one, you can see Eckbo's burst of creative energy. The spatial logic of his beaux-arts training is still evident, but it has exploded into a bombastic display of free-form, non-linear landscape space. A new style of rendering emerged that reflected his influences from jazz and cubism. To express these revolutionary new forms of the California Garden Style, he used a #2 watercolour brush to creating the sensuous, sweeping forms.

Birds-Eye Axonometric Garden Views

Refer to Figure 9.6a, Cole Garden, Oakland, CA, 1941. The axonometric view is an excellent drawing convention to represent the overall relationship between the building interior and the garden. This technique became the dominant method to illustrate the California Garden Style at mid-century. Since the sizes of objects do not diminish into the distance as in a perspective rendering, the axonometric is an ideal way to show how vegetation and built structures work together to define the garden space.

Refer to Figure 9.6b, Garden in Woodside, CA, 1947. The solid black shape of the building plan anchors the drawing to the bottom and left side of the page. Thick, loopy brush strokes illustrate the tree canopies. A variety of vertical and horizontal hachures define trees and hedges, and a gradient of stipples creates the grass texture on the ground plane.

Refer to Figure 9.6c, Elementary School in Culver City, CA, 1948. This visually powerful composition is accented by a line of tall, sensuously inscribed trees in the background, and amoeba-shaped trees in the foreground. A zig-zag hedge is drawn with tight squiggly lines to define its sides, and the planter beds are filled with tiny circles and loopy textures to describe the ground planes. We also see an early use of Zip-A-Tone, a thin textured film used to highlight the water in the central pool.

"Quazy" Atomic Plants

Refer to Figure 9.7a, Garden in Holmby Hills, Los Angeles, CA, 1947. In my own practice, I have tried to recreate the style of drawing vegetation from the mid-century modernists only to fail miserably. This elevation expresses an incredible abstraction of plant forms drawn with pen, brush, and ink. Heavy stipples define the ground planes. The brush helps create flowing, amoeba-like plant forms. Thin ink lines form the interior branching structure, and black dots add texture to the tree canopies.

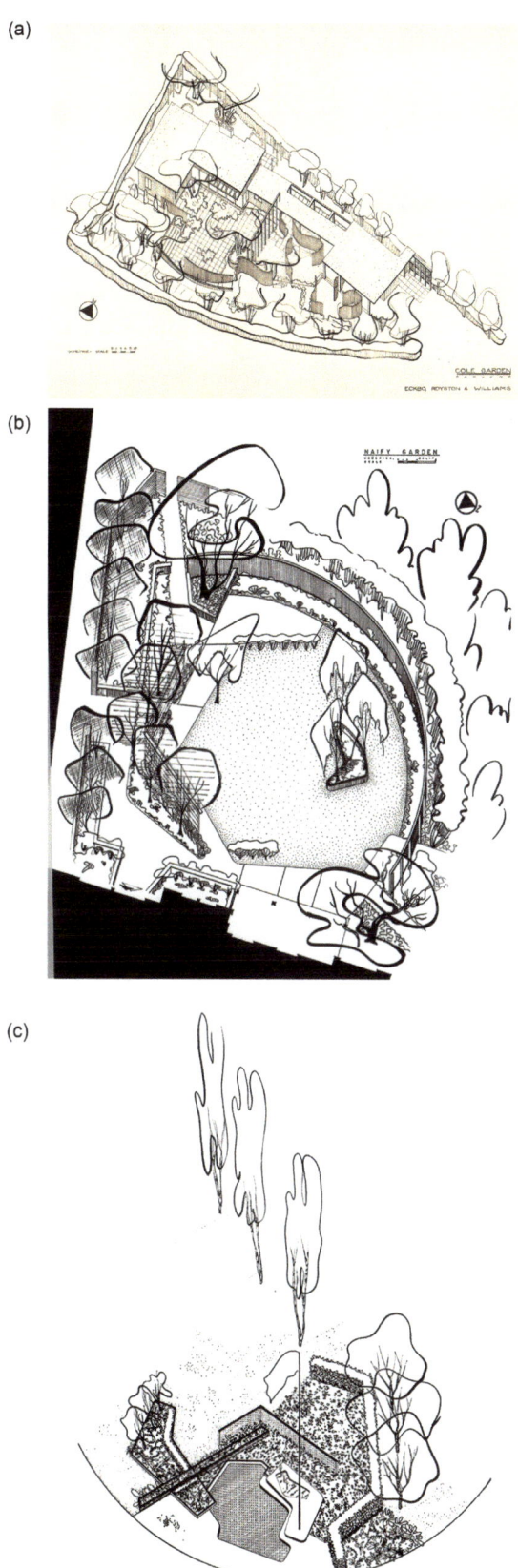

9.6
(a) Cole Garden, Oakland, CA, 1941. By Garrett Eckbo. Environmental Design Archives, University of California Berkeley. (b) Garden in Woodside, CA, 1947. By Garrett Eckbo. Environmental Design Archives, University of California Berkeley. (c) Elementary School in Culver City, CA, 1948. By Garrett Eckbo. Environmental Design Archives, University of California Berkeley.

9.7
(a) Garden in Holmby Hills, Los Angeles, CA, 1947. By Garrett Eckbo. Environmental Design Archives, University of California Berkeley. (b) Mineral King Co-operative Ranch, Community Park, CA, 1939. By Garrett Eckbo. Environmental Design Archives, University of California Berkeley.

Refer to Figure 9.7b, Mineral King Co-operative Ranch, Community Park, CA, 1939. This drawing lies somewhere between an axonometric view and a bird's eye perspective. It is a slightly distorted aerial perspective showing a comprehensive view of the space. The wacky vocabulary of plant forms seems to me almost like a botanical zoo! The three-dimensional volumes of the vegetation are created with dense, expressive leaf textures and abstracted branching structures.

Refer to Figure 9.8, The California Garden Series. When I arrived in California to begin teaching at UC Berkeley, I wanted to experiment with the design language of the California Garden Style and explore how the next generation of garden forms might relate to environmental conditions at the time. My theoretical gardens were inspired by three major forces: the drawings of Garrett Eckbo and Thomas Church, the cubist gardens exhibited at the 1925 Art Deco exposition, and the spontaneous prose of Jack Kerouac based on jazz improvisation. My gardens are laid out with a strong formal geometry, but then break apart into asymmetrical spaces. The visual planes constantly shift, distorting

9.8
The California Garden Series. By Chip Sullivan.

the perspective. In this series, the hidden flows of nature are brought to life through sweeping interconnected lines, patterns, strong colour orchestration, and brilliant contrasts.

These explorations were created on 10 inch × 14 inch Arches 140 lb. cold press watercolour paper, inked with a Speedball #512 bowl point and crow quill dip pen. Colour was applied with Winsor & Newton Series 7 sable hair brushes #4 and #10.

Summary

From the late 1940s to the mid-1950s, Church and Eckbo were the avant-garde of garden design. The California Garden Movement was one of the last significant garden styles in landscape history. Landscape historian David Streatfield describes the legacy of Eckbo and Church as "one of the most significant contributions to landscape design since the Olmstedian tradition of environmental planning in the second half of the 19th century."[1] I strongly believe that the drawing media used in the creative process and ultimate rendering of these designs had a profound effect on the garden style that emerged. Their beaux-arts training provided Church and Eckbo with solid drawing and compositional skills, while the tenets of Modernism gave them freedom to explore totally new forms of expression. Tommy Church and Garrett Eckbo were fortunate to have developed skills that built on influences from both the beaux-arts and Bauhaus educational systems. Like many design innovators of the mid-twentieth century, their careers formed a bridge between eras.

To see the sensuous and elegant hand-lettering on these drawings is magical. Eckbo's gothic-style lettering on his early beaux-arts style drawings and Church's personal touch in his pencil lettering makes me appreciate the skill and craft involved in creating a finished work of art. The invasion and dominance of Chartpak rub-on Helvetica type that soon followed this period virtually wiped out the beauty and artistry of hand-lettering.

Oddly enough, there are not a lot of human figures in renderings from this era. The few figures that are represented resemble the misshapen Corbusier Modulor man or the clay animation figure Gumby. There too is an odd similarity to Henry Moore's sculptures expressive of post-war existentialism and fear of the Atomic Age.

Both Eckbo and Church realized the importance of promoting landscape architecture through the publication of articles, books, and museum exhibitions. The museum exhibitions helped establish landscape architecture as an art form and exposed the California Garden Style to a larger public audience and critical review. Eckbo and Church made extensive use of models and relief drawings in their exhibitions. The three-dimensional medium in a museum context made it much easier for the viewing public to see landscape architecture as a form of sculpture.

The drawing tools and design methods we employ certainly impact the quality and style of our art. Eckbo and Church helped advance a unique period in the history of landscape architecture representation. In the following decades, we saw the arrival of magic markers—what Louis Kahn referred to as "those vulgar pencils that smell like benzene"[2]—and the predominance of Zip-A-Tone, Chartpak AD Markers, and rubber stamps. Hopefully, the creative rendering techniques and methods described in this book will revive the art of landscape representation and lay the foundations for a renaissance of landscape drawing.

Notes

1 Streatfield, D. "Where Pine and Palm Meet: The California Garden as Regional Expression." *Landscape Journal*, 1986, 4(2): 61.
2 Solomon, D. "Fixing Suburbia." In: A. Weisser (Ed.). *The Pedestrian Pocket Book: A New Suburban Design Strategy*. New York, NY: Princeton Architectural Press, 1989, p. 30.

References

Barr, A. *Cubism and Abstract Art*. New York, NY: Museum of Modern Art, 1936.
Church, T. *Gardens Are for People*. New York, NY: Reinhold, 1955.
Contemporary Landscape Architecture and Its Sources. San Francisco: Museum of Modern Art, 1937.
Eckbo, G. *Landscapes for Living*. New York, NY: Duell, Sloan, and Pearce, 1950.
Eckbo, G. "The Eesthetics of Planting." In: *Landscape Design*. San Francisco, CA: San Francisco Museum of Art, 1948, pp. 16–18.
Exposition Internationale des Arts Decoratifs Moderne. Paris, France: H. Tourte et M. Petitin, 1925.
Hubbard, H.V. and Kimble, T. *Introduction to the Study of Landscape Design*. New York, NY: Macmillan, 1917.
Imbert, D. *The Modernist Garden in France*. New Haven, CT: Yale University Press, 1993.
Landscape Design. Exhibition Catalogue. San Francisco, CA: San Francisco Museum of Art, 1949.
Trieb, M. (ed.). *Modern Landscape Architecture: A Critical Review*. Cambridge, MA: The MIT Press, 1993.
Treib, M. (ed.). *Thomas Church Landscape Architect: Designing a Modern California Landscape*. San Francisco, CA: William Stout Publishers, 2003.
Treib, M. and Imbert, D. *Garrett Eckbo, Modern Landscapes for Living*. Berkeley, CA: University of California Press, 1997.
Solomon, D. "Fixing Suburbia." In: A. Weisser (Ed.). *The Pedestrian Pocket Book: A New Suburban Design Strategy*. New York, NY: Princeton Architectural Press, 1989, p. 30.
Streatfield, D. "Where Pine and Palm Meet: The California Garden as Regional Expression." *Landscape Journal*, 4(2) (1986): 61.

10 The Drawings of Lawrence Halprin

Alison Hirsch

Lawrence Halprin (1916–2009) was one of the most prolific landscape architects of the twentieth century, particularly during the period of significant urban change in the 1960s and 1970s United States. In addition to nationally notable built projects in cities undergoing dramatic "urban renewal" (Seattle, Denver, San Francisco, Portland, etc.), Halprin's extensive writings, including the densely illustrated books *Cities* (1963, revised 1972), *Freeways* (1966), *The RSVP Cycles* (1970) and *Taking Part* (1974, with Jim Burns), offer insight into both the practice and theory behind his graphic experiments, especially as they attempted to represent forces of change characteristic of the time.

Prior to working on large-scale urban projects through Lawrence Halprin & Associates (established in 1960), Halprin began his own post-war practice by designing private gardens in California. The garden provided a laboratory to experiment with spatial sequence, texture, and tactility, growth, and change as well as provided experience in notating these dynamics graphically (Figure 10.1). In one article appearing in the annual dance magazine, *Impulse*, he describes his conception of the garden's evolution in the booming post-war U.S., "We are no longer content to sit stiffly in the garden in our best Sunday clothes… Our gardens have become more dynamic and should be designed with the moving person in mind."[1] Rather than treat gardens—and later cities—as spatial problems to be fixed in master plans, Halprin instead conceived of them as spatio-temporal systems that could be structured to stimulate heightened human creativity and socialization. However, to do so, techniques had to be developed to visualize movement and change.

Ecology

Halprin spent much of his career experimenting with visual languages that he hoped could be used to adequately address the dynamics of modern life. He conducted these early experiments studying not the transformations of the city, but their precedent—the "archetypal" processes of growth, disturbance, and entropy he discovered in the High Sierra and dramatic coast of Northern California. He studied natural process in the academy and in the field, obsessively observing his surroundings through repeated sketching, recognizing that ecosystems did not necessarily behave in a way that was consistent with the equilibrium paradigm—the idea that all ecosystems are moving toward a stable state of climax or balance. Largely pre-empting a shift in ecology from the study of ecosystems as closed, self-regulating systems to ecosystems as intrinsically open to frequent disturbance, Halprin found the vocabulary of disturbance and "disclimax" applicable to the dramatically shifting urban landscape. Just as species and systems exhibited a wide range of adaptations to disturbance, Halprin aimed to guide humans through their adaptation to a "whole new set of conditions."[2]

Halprin's visual studies of the Sierra, Nevada, and the rugged coast of Northern California became aesthetic sources for this urban work. While sketching, he focused not on the scenic qualities of these landscapes or the compositional codes of drawing, but attempted to represent processes that shaped the world around him. Experimenting with different modes of mark-making combined with words, evoking sounds and actions, and temporal notations, these drawings served as the foundation for his design process (Figures 10.2 and 10.3). Believing in the binding power of humans' common

DOI: 10.4324/9781003183402-10

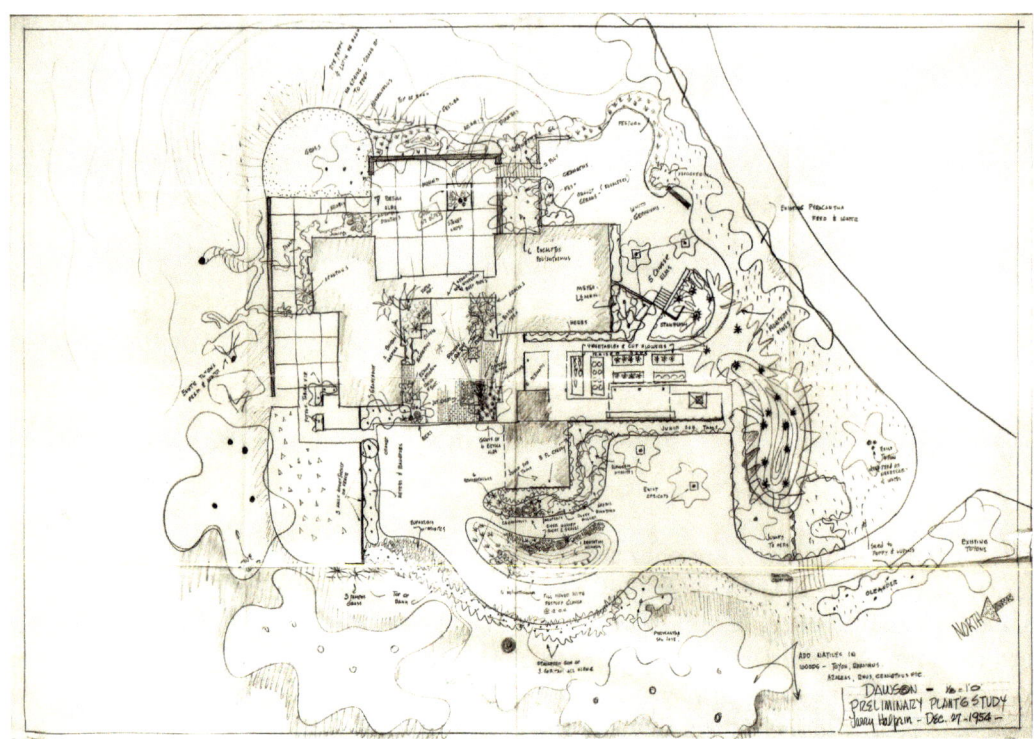

10.1
Preliminary Planting Study for the Dawson Garden, Los Altos, California, 1953–1955. Halprin draws primarily in plan while some of the plantings are pictured in perspective and elevation. The Lawrence Halprin Collection, The Architectural Archives, University of Pennsylvania (014.II.A.37).

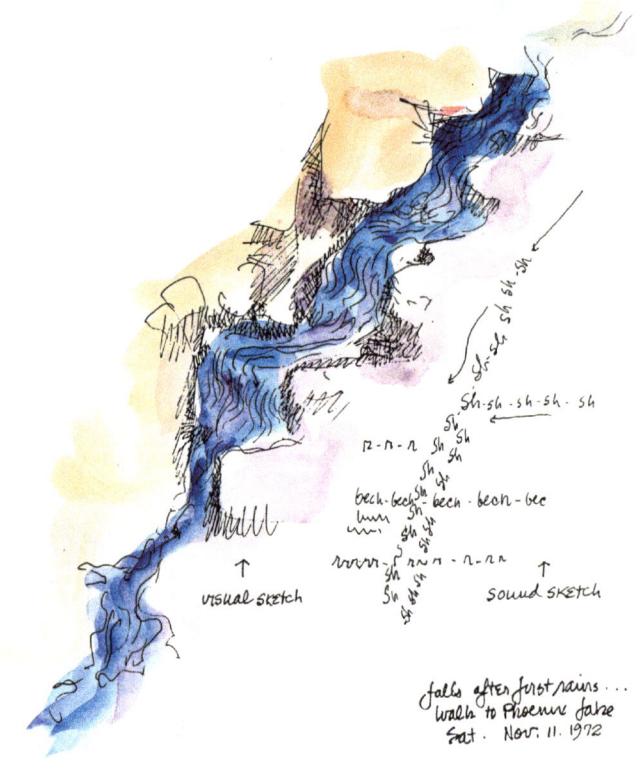

10.2
Falls near Phoenix Lake, Marin County, 1972. The Lawrence Halprin Collection, The Architectural Archives, University of Pennsylvania.

10.3
High Sierra sketches, 1964. The Lawrence Halprin Collection, The Architectural Archives, University of Pennsylvania.

origins in nature and natural forces of creation, Halprin used this "archetypal" vocabulary of processes such as erosion, deposition, and plant succession to resonate broadly through the fiercely polarized social landscape of the 1960s American city.

Choreography

While Halprin's early academic studies of the natural sciences inspired his process-driven practice, his dedication to the temporal dimension was exponentially compounded by the artistic symbiosis that existed between him and his wife, avant-garde dancer and choreographer Anna Halprin. It was thus the combination of ecology and dance that motivated his lifelong search for methods of representation that recorded, "transmuted" (a word he liked), and propelled open-ended process and change (Figure 10.4).

Like "Happenings," emerging from the teachings of musician John Cage in New York, Anna organized interactive events in which environmental situations and loose action guidelines were proposed or "scored," but the ultimate performance was left open-ended and typically involved the audience. From these new art forms, the "open score" became the major tool for stimulating action and transforming the spectator from observer to participant. Lawrence Halprin integrated these emerging performance theories into his creative process, both in drawing—through graphic "scores" that recorded and choreographed movement (Figure 10.5)—and in building—designing spaces as "scores" intended to stimulate open-ended participation and kinaesthetic response.

Motation

In a 1965 article appearing in Progressive Architecture, he proposed a notational system called "Motation" (combining "movement" and "notation") as a supplement to conventional architectural drawings, such as plans and elevations, which he considered too static.[3] In filmic sequence, the frames

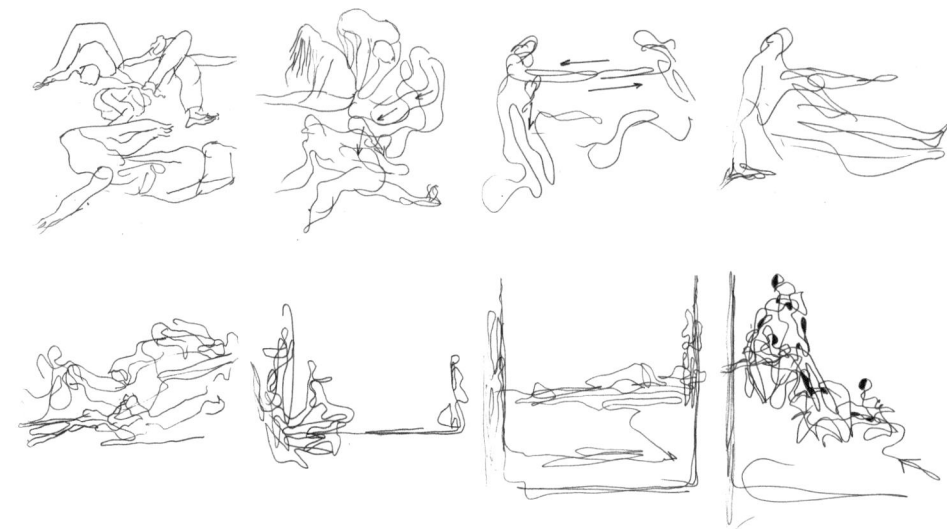

10.4
Lawrence Halprin's sketches of Anna Halprin's workshops on the dance deck, 1966. The Lawrence Halprin Collection, The Architectural Archives, University of Pennsylvania.

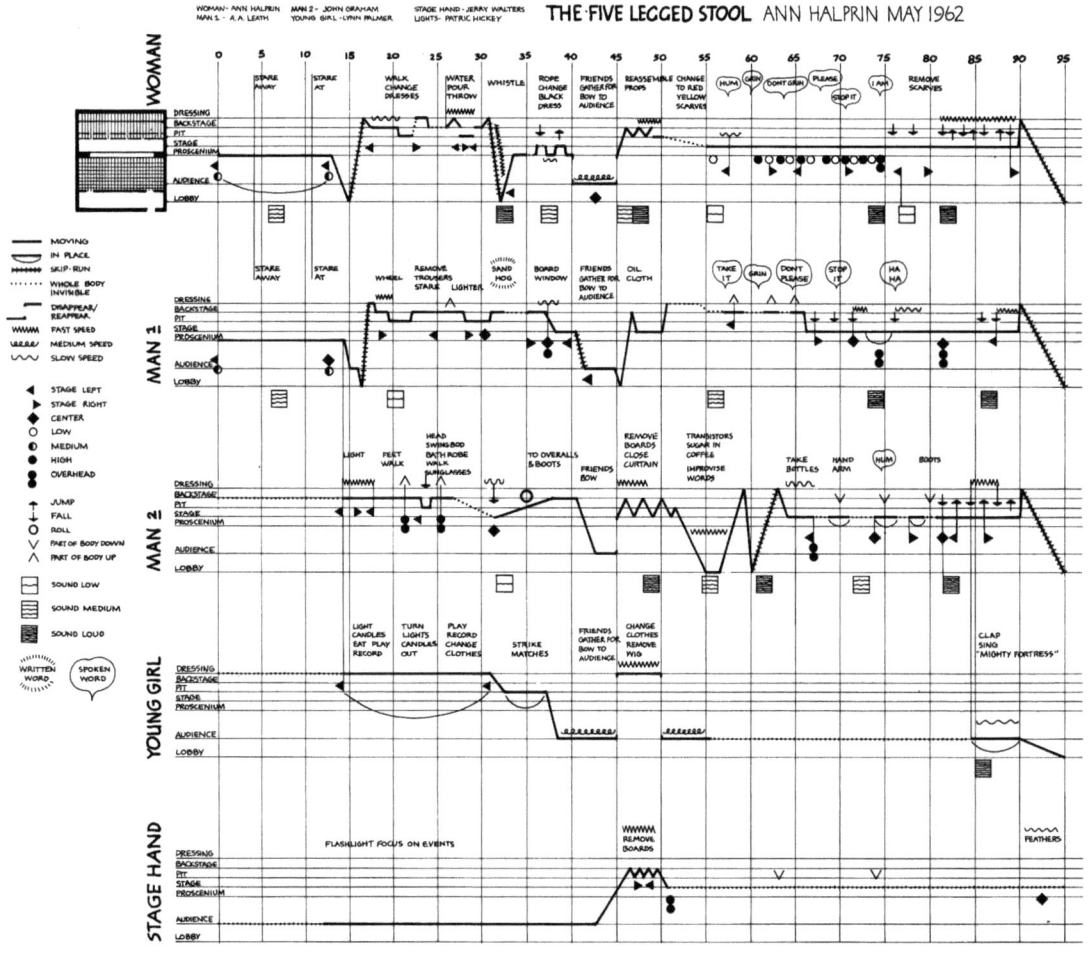

10.5
Lawrence Halprin's graphic score for Anna Halprin's performance of The Five-Legged Stool, May 1962. The Lawrence Halprin Collection, The Architectural Archives, University of Pennsylvania.

The Drawings of Lawrence Halprin 81

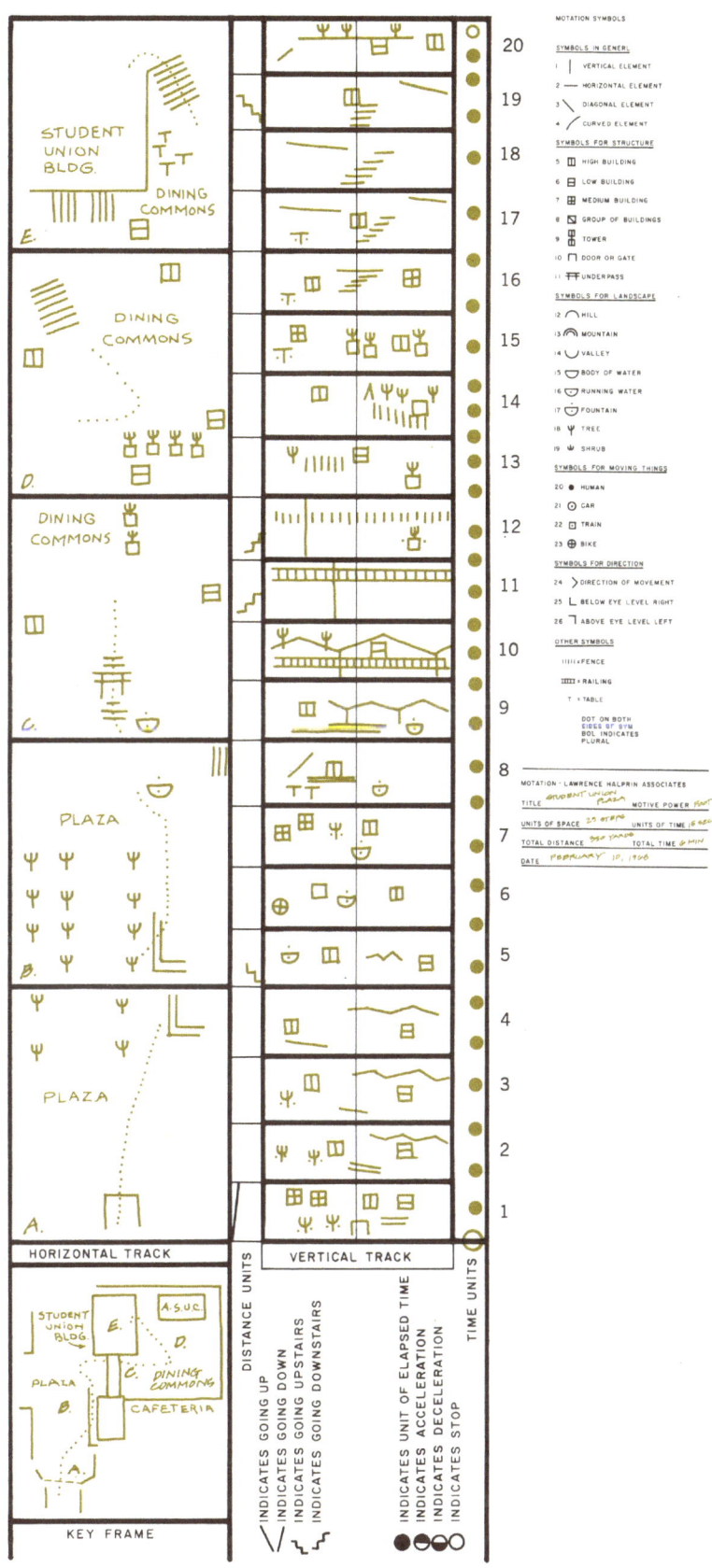

10.6
"Motation" study for Student Union Plaza, University of California, Berkeley, 1966. The Lawrence Halprin Collection, The Architectural Archives, University of Pennsylvania.

of "Motation" provide both a horizontal (plan) and vertical (elevation) understanding of spatial relationships and durational measure (Figure 10.6). By distilling the visual world into a system of symbols, rhythms could be extracted and set in comparison or be used productively to choreograph shifts in the everyday repetitions of life. Halprin's fascination with symbolic communication paralleled his search for archetypal imagery, as he attempted to create a universal vocabulary for movement (Figure 10.7).

Rather than merely a descriptive recording device, which Halprin considered the function of Labanotation,[4] he became most interested in how a designer might represent movement as a starting

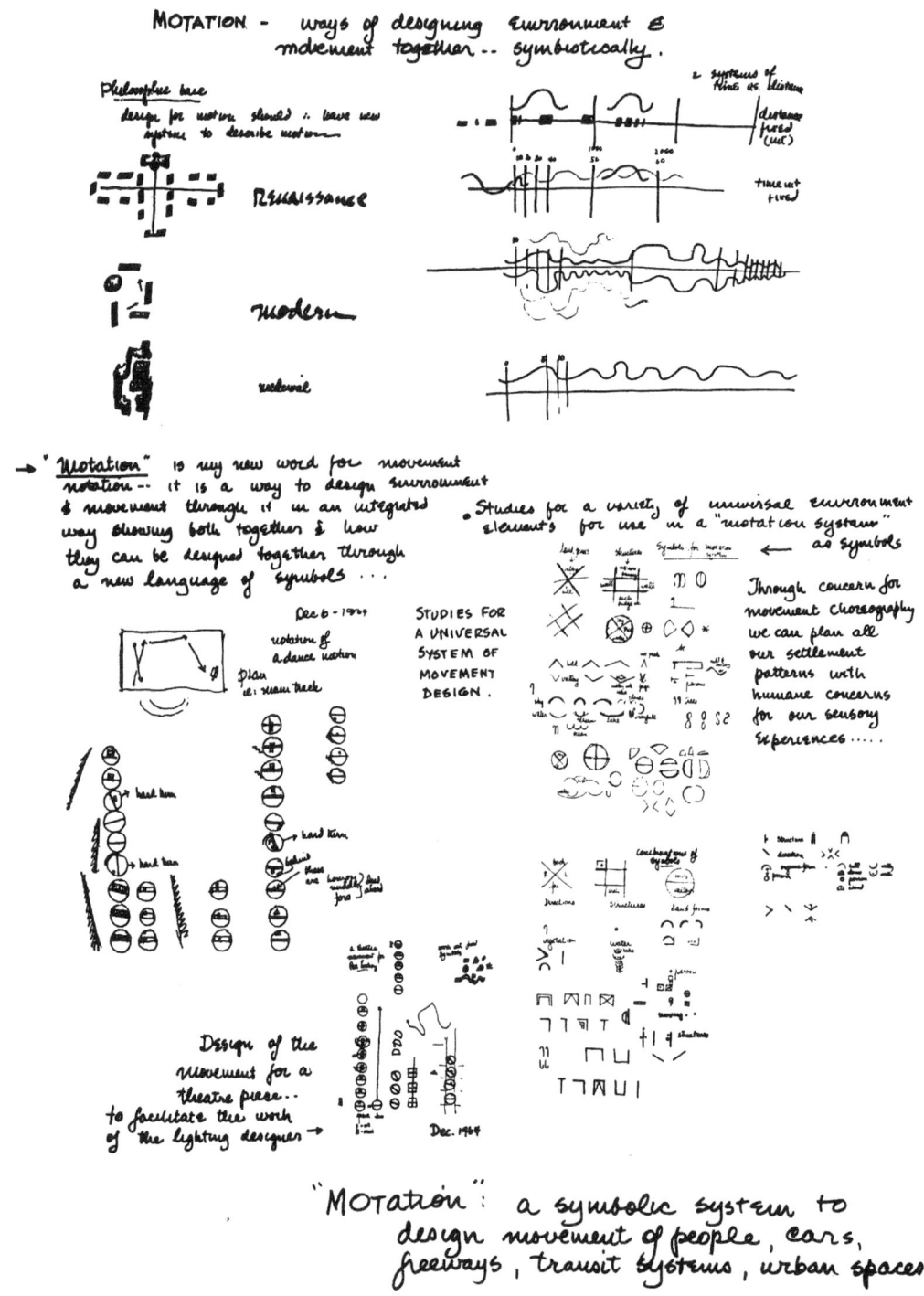

10.7
Composite notes for "Motation" from sketches in the early 1960s. Printed in Sketchbooks of Lawrence Halprin. Tokyo, Japan: Process Architecture, Co., 1981, p. 29. The Lawrence Halprin Collection, The Architectural Archives, University of Pennsylvania.

point from which to generate form. In his article on "Motation," he states, "Only after programming the movement and graphically expressing it should the environment—an envelope within which movement takes place—be designed. The environment exists for the purpose of movement."[5]

More specifically, "Motation" uses a series of "frames" that Halprin compares to motion picture film to depict sequence or linear progression through space. First, the plan of the overall voyage is drawn in the "Key Frame" on the lower left of the "Motation Form" Halprin has standardized (refer to Figure 10.6). The "Horizontal Track" on the left side is a plan representation of the journey broken into a series of frames. Corresponding to those frames, the Vertical Track includes "the 'normal' visual horizon," presenting the third dimension of movement experience as a series of framed sequential views. He provides smaller spaces for the symbolization of other sensorial events such as sound, smell, colour, or rain in the "Distance Strip" between the Horizontal and Vertical Tracks. The Distance Strip additionally indicates how far one has travelled as well as the rise and fall of the terrain through specific symbols. To the right of the Vertical Track, the Time Strip provides space for recording tempo through the use of dots appearing closer together (fast) or farther apart (slow).

In practical application, "Motation," rather than beginning with abstract movement to generate form, really became a "testing-out device" for Halprin in the early 1970s. He claims that these notations "allow us to see—in our mind's eye—possible changes in the environment before actually carrying them out. They are a way to test our innumerable alternate futures."[6] Though Halprin did not employ a rigorous "Motation" process in the designs for his own urban environments, he proposed it as a tool for design pedagogy and practice.[7]

Water

Water became a major choreographic force that guided sensorial and kinetic response through his designs. Halprin spent his life obsessively sketching the behaviours of water in dramatic settings such as Sea Ranch and the High Sierra (Figure 10.8). From these direct observations, he designed complex

10.8
Wave action off the coast of Sea Ranch during a storm, 1967. The Lawrence Halprin Collection, The Architectural Archives, University of Pennsylvania.

water features throughout his career that maximized on the element's variable natures. Water could be scored to stimulate movement; curiously finding its source, seeking out its sound, playing in its mist and sprays, and splashing in its pools were human responses that Halprin consciously choreographed, believing in water's universal resonance as an elemental force and "atavistic need."[8]

Lovejoy Plaza and Ira Keller Fountain in Portland, as well as the Canyon Fountain in Seattle Freeway Park, provide excellent examples of how he translated his coastal Northern California and Sierra studies into dynamic fountains intended not to provide "respite" from dramatically shifting urban contexts but match the energy of movement and change characteristic of the time and place. Graphically, Halprin designed scores for the "performance" of a series of such interactive fountains (Figures 10.9 and 10.10).

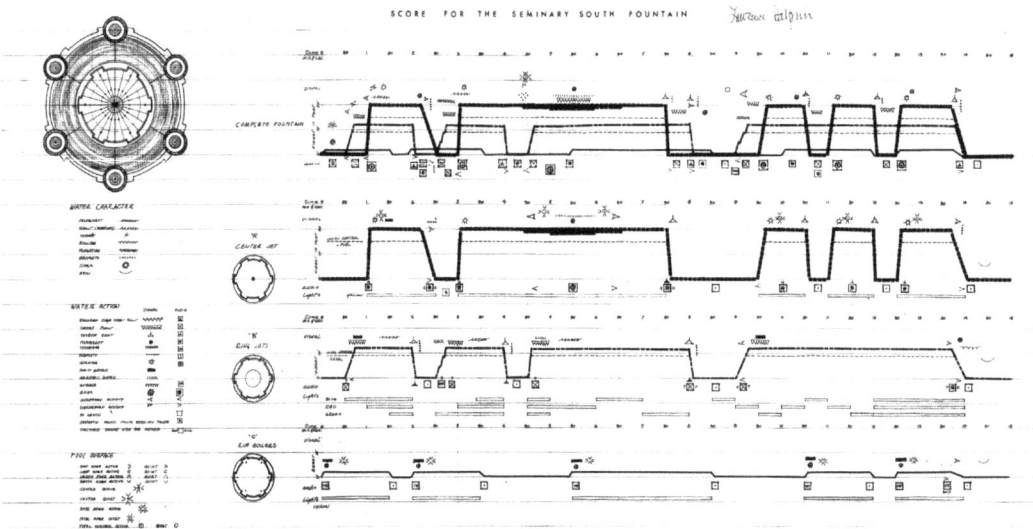

10.9
Score for the Seminary South Shopping Center Fountain, Fort Worth. In Cities (1972 edition), Halprin describes: "Each group of jets is drawn as a graph separately, and the height is plotted against time. Under each are the secondary elements notated with specially designed symbols which indicate sound, surface characteristics, shape of droplets, etc. Finally, the various jet groupings are combined together in the top graph as a means of relating them all together in orchestral effect" (p. 160). The Lawrence Halprin Collection, The Architectural Archives, University of Pennsylvania.

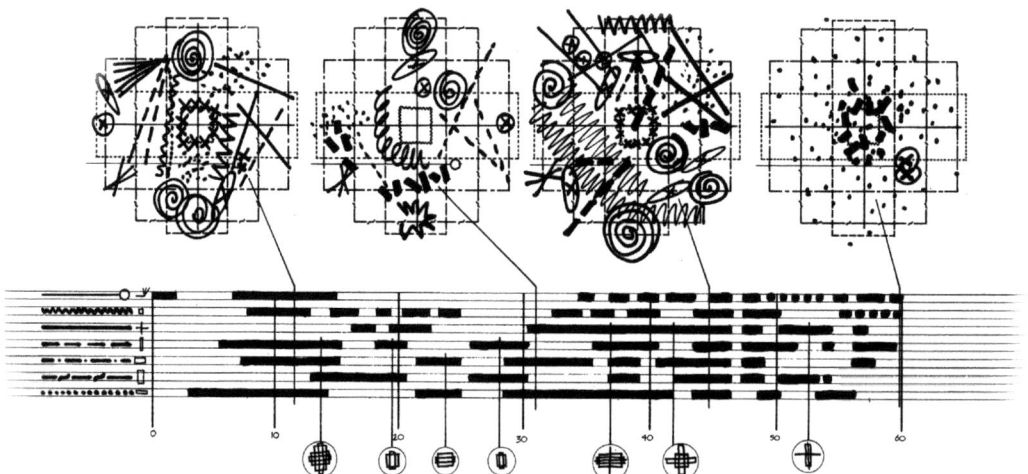

10.10
Score for the Seattle Center Fountain designed with Jacques Overhoff, early 1960s. The jets were designed to respond to counterpressures of atmospheric conditions such as wind. The Lawrence Halprin Collection, The Architectural Archives, University of Pennsylvania.

These spatio-temporal graphic experiments provided a foundation and stimulus for the development of representational methods that record, generate, and realize change. Though Halprin's creative points of origin were open-endedness, process, movement, and change, the translation from observation, to notation, to application in built form ultimately endowed the city with strong material presence that served as a shared framework for creative appropriation. Halprin responded to the powerful forces of modernization that shaped the city with equally forceful spaces that counteracted alienation from the natural world through the reintroduction of rhythms and natural forces of change abstracted in concrete. His experiments in visualization and his desire to provide both flexibility of appropriation and strong material presence may serve as an example for designers shaping a world whose pace of change only accelerates.

Notes

1. Halprin, L. "The Choreography of Gardens. *Impulse: Annual of Contemporary Dance* (1949): 32-33.
2. Lecture by Halprin, L. "The Human Community as an Ecosystem." AIA Northwest Regional Conference in Tacoma on September 9, 1963 (Lawrence Halprin Collection, University of Pennsylvania Weitzman School of Design Architecture Archives, Philadelphia, PA, 014.I.A.6035).
3. Halprin, L. "Motation." *Progressive Architecture*, 1965, 46(7): 126–133.
4. Rudolf Laban was known for his unique use of the vertical staff to represent the body. Laban's system is limited to the movement of the body—its direction, timing, and simultaneous gestures of its many parts, not to convey the body's shifting relationship to the environment, which was one primary goal of Halprin's Motation.
5. Halprin, L. "Motation." *Op. cit.*, p. 126.
6. See "Calligraphy as a Structuring Device" essay draft (Lawrence Halprin Collection, University of Pennsylvania Weitzman School of Design Architecture Archives, Philadelphia, PA, 014.I.B.2479).
7. In a 1966 *Progressive Architecture* article entitled "Urbanography," the journal editors present research on the "notation of sequential experience in cities" by faculty within the architecture department of the University of Cincinnati. Two assignments by Professors Abernathy, Noe, and Goetzman demonstrate the further development of the notational schemes of Halprin, Kevin Lynch, Philip Thiel and others, to accommodate a full range of sensory stimuli and to use the process and symbolic language to *design* spatial sequences (*Progressive Architecture*, 1966, 47(4): 184–190).
8. Halprin, L. *Cities*. New York, NY: Reinhold Publishing Corporation, 1972, p. 134.

References

Abernathy, B.L. and Noe, S. "Urbanography." *Progressive Architecture*, 47(4) (1966): 184–190.
Halprin, L. "Motation." *Progressive Architecture*, 46(7) (1965): 126–133.
Halprin, L. "The Choreography of Gardens." *Impulse*, 1949: 32–33.
Halprin, L. *Cities*. New York, NY: Reinhold Publishing Corporation, 1972.
Lawrence Halprin Collection, University of Pennsylvania Weitzman School of Design Architecture Archives, Philadelphia, PA.

11 Ian L. McHarg and Mapping Complex Processes

Frederick R. Steiner

Ian McHarg opened a new way for us to see the world. His approach for interpreting the play between natural and cultural systems has become the dominant visualization technology of our time, just as Filippo Brunelleschi's experiment with linear perspective dominated architectural visualization for over 600 years. Ian McHarg provided a roadmap for applying ecological information to how we interpret, plan, and shape our surroundings. This became his quest, his principal contribution. Whereas Brunelleschi did not exactly invent perspective, he did establish a system of linear perspective around 1,420. Similarly, map overlays existed before McHarg, back at least to the Olmsted office in the 1890s. McHarg created a method for overlaying maps of biophysical and sociocultural information grounded in ecological knowledge.

McHarg developed his method on a solid bedrock of work in overlays by other landscape architects and planners, but McHarg's contributions were unique and important. Landscape architects began using hand-drawn, sieve-mapping overlays in the late nineteenth century. Charles Eliot and his associates in the office of Olmsted, Olmsted, and Eliot pioneered overlays through sun-prints produced on their office windows. McHarg acknowledged the early contributions of Charles Eliot, an Olmsted protégé, who worked systematically with scientists to collect and map information to be used in planning and design. Warren Manning, an apprentice of both Olmsted and Eliot, used soil and vegetation information with topography and their combined relationship to land use to prepare four different maps of the town of Billerica, Massachusetts, in 1912. Manning's Billerica Plan displayed recommendations and changes in the town's circulation routes and land use.[1]

Eliot left the most explicit explanation about why and how the overlays were employed. After Eliot's death at 37, his father, Charles W. Eliot, the president of Harvard, published the biography *Charles Eliot, Landscape Architect*,[2] which was an interpretation of his son's work. This book provides perhaps the first account of the overlay technique. The Boston Metropolitan Park work undertaken by Olmsted, Olmsted, and Eliot, with Eliot in charge, involved six months of "diligent researches."[3] Eliot used a variety of consultants, including a Massachusetts Institute of Technology professor, as well as Olmsted staff members, to conduct surveys of the metropolitan region's geology, topography, and vegetation. These maps provided the basis for the overlay process, which Eliot describes as follows:

> By making use of sun-prints of the recorded boundary plans, by measuring compass lines along the numerous woodpaths, and by sketching the outlines of swamps, clearings, ponds, hills, and valleys, extremely serviceable maps were soon produced. The draughting of the several sheets was done in our office. Upon one sheet of tracing-cloth were drawn the boundaries, the roads and paths, and the lettering … on another sheet were drawn the streams, ponds and swamps; and on a third the hill shading was roughly indicated by pen and pencil. Gray sun-prints obtained from the three sheets superimposed in the printing frame, when mounted on cloth, served very well for all purposes of study. Photo-lithographed in three colours, namely, black, blue, and brown, the same sheets will serve as guide maps for the use of the public and the illustration of reports.

> Equipped with these maps, we have made good progress, as before remarked, in familiarizing ourselves with the lay of the land.[4]

After Eliot and Manning, there were several studies in which the use of the overlay technique is apparent, but a theoretical explanation about the rationale for using the technique as an orderly planning method was missing. The city plan for Dusseldorf, Germany, in 1912; the Doncaster, England, regional plan in 1922;[5] the 1929 regional plan of New York and its environment;[6] and the 1943 London County plan[7] incorporate typical characteristics of the overlay process.[8] Thomas Adams, who directed the extensive 1929 New York regional planning study, addressed suitability in his 1934 *The Design of Residential Areas*, but mostly from an economic perspective. An academic discussion of the overlay technique did not surface until 1950 and publication of the *Town and Country Planning Textbook* containing a contribution by Jacqueline Tyrwhitt that dealt explicitly with the overlay technique.[9]

In an example given by Tyrwhitt,[10] four maps (relief, hydrology, rock types, and soil drainage) were drawn on transparent papers at the same scale and referenced to common control features. These data maps were then combined into one land characteristics map that provided a synthesis, an interpretation, and a judicious blending of the first four maps.[11] This sieve-mapping overlay method was widely accepted and incorporated in the large-scale planning of the British new towns and other development projects after World War II.[12] Before the war ended, Ian McHarg took a correspondence course offered by Tyrwhitt and others. McHarg was also involved in new town planning in the early 1950s in Scotland, so he was quite aware of the British new town program. As a result, he was introduced to the concept of overlay mapping early in his career.

George Angus Hills' plan for Ontario Province[13] is a pioneering North American example that employed a well-documented data-overlay technique.[14] Hills was on the staff of the Ontario Department of Lands and Forests. His technique divides regions into consecutively smaller units of physiographic similarity based on a gradient scale of climate and landform features. Through a process of comparing each physiographic site type or homogeneous land unit to a predetermined set of general land-use categories and rankings of potential or limitation for each use or activity, the resulting units were regrouped into larger geographic patterns called 'landscape units' and again ranked to determine their relative potentials for dominant and multiple uses. The land-use activity with the highest feasibility ranking within a landscape unit was recommended as a major use.[15] The Hills method has been influential in the development of the Canadian Land Inventory System.[16]

The year after Hills' Ontario plan, Philip Lewis, a landscape architect at the University of Wisconsin–Madison and principal consultant to the Wisconsin Department of Resource Development, applied an overlay analysis technique to evaluate natural resources for the entire state of Wisconsin.[17] Lewis had been recruited to Wisconsin by Gaylord Nelson to start environmental planning in the state. Lewis' work was a direct response to the growth and demand for outdoor recreation across Wisconsin. According to Forster Ndubisi, "Unlike Hills, whose work was based primarily on examining biological and physical (biophysical) systems such as landforms and soils, Lewis was concerned more with perceptual features such as vegetation and outstanding scenery."[18] Lewis stressed the importance of the patterns, both natural and cultural, within the landscape. He combined individual landscape elements of water, wetlands, vegetation, and significant topography through overlays onto a composite map depicting Wisconsin's areas of prime environmental importance.[19] By combining resource inventory data and soil survey data, Lewis was able to create maps that identified intrinsic (natural) patterns. Once additional resources were grouped by patterns and mapped, points were assigned to major and additional resources and totalled to identify relative priority areas. Demand for planned uses and limitations of each priority area for specific uses were then combined to assign specific uses to each priority area.[20]

McHarg, Lewis, and Hills refined their approaches during the 1960s and built their work on all these earlier efforts. McHarg especially advanced previous methods significantly by linking suitability analysis with theory. He provided a theoretical basis for overlaying information. McHarg's approach focused on mapping information on both natural and human-made attributes of the study area, initially as individual transparent maps.[21] Each map layer was hand-made using a combination of drawing and a type of coloured (or toned) plastic film (of a self-adhesive variety sold under brand names like Pantone or Zip-A-Tone). The "support" for these sheets was likely clear acetate film. Based on surviving evidence in the University of Pennsylvania Architectural Archives (the Philadelphia health study—which shows careful registration marks at sheet edges), the various layers to be "activated" were then run through a diazo-type machine (Elliot's "sun prints," our "blueprints"). This made the composite X-ray-like representation. The result was a diazo-type blueprint. The composite maps illustrated intrinsic suitabilities for land-use classifications such as conservation, urbanization, and recreation for the specific planning area. These maps were then combined with each other as overlays to produce an overall composite suitability map.[22]

McHarg's inventory process provides one of the first examples of methodological documentation for the overlay technique (with those by Hills and Lewis). McHarg was also the first to advocate the use of the overlay technique to gain an ecological understanding of a place. He noted that

> a region is understood as a biophysical and social process comprehensible through the operation of [natural] laws and time. This can be reinterpreted as having explicit opportunities and constraints for any specific human use. A survey will reveal the most fit locations and processes.[23]

As a result, he was explicit about the range and sequence of mapped information to be collected. McHarg also observed that the phenomena represented by the maps were valued differently by various groups of people and thus could be weighted differently, depending on the circumstance.

Maps as Slices of Time

McHarg used the analogy of a layer cake to explain his approach for collecting and making mapped information about the biophysical and sociocultural processes of a place (Figure 11.1). This information could be collected from the site to the regional scale building up layers of maps. McHarg advocated that it should be collected at the national and global scales, a vision that has happened as computer mapping and satellite imaging has progressed. The biophysical layers included climate, geology, physiography, ground and surface water hydrology, soils, plants, animals, and land use. He also suggested various sociocultural maps such as settlement history, current land use, ethnography, demographics, and land value. Each map represented a process, a slice in time, based on the best existing science (Figure 11.2). McHarg recognized that science is dynamic and changing but that design and planning should be based on the best existing data. To this end, he often enlisted local experts in geology, hydrology, soils science, limnology, and biology to complement the skills of the landscape architects and planners in his firm (Wallace, McHarg, Roberts, and Todd). Likewise, he engaged similar scientists as well as ethnographers and economists in his studios at the University of Pennsylvania (Penn) Department of Landscape Architecture and Regional Planning.

For McHarg, the maps were one component—an important one—of a larger project to understand landscapes and to help make wise choices about their futures. For instance, for nine years, in 1960–1961 and then from 1966 to 1974, McHarg focused his Penn studios on the Delaware River Basin. The students produced massive acetate maps of the basin's layers (Figure 11.3). McHarg and his students and faculty used light tables instead of office windows to draw these maps colouring them with pencils and magic markers. McHarg's protégé Narendra Juneja emerged as a master of the use of colours on these maps. As the flow chart (Figure 11.4) from his 1967 studio illustrates, the production of resource maps was a step in a process to help map health, generate options, and determine suitabilities for various land uses.

McHarg's ideas about what a place was suitable for was driven by his concerns about human health and safety and informed by Darwinian fitness as refined by Lawrence Henderson. Each place possesses a range of opportunities and constraints. McHarg contended that, by overlaying maps, intrinsic suitabilities would be revealed where some uses fit better than others. The process would also reveal who suffered and who benefited from land-use decisions. McHarg's method can be employed for both development and conservation.

Of course, this approach has been advanced considerably by geographic information systems (GIS) technologies. McHarg began experimenting computer mapping in the 1960s. In his Fall 1968 and 1969 Delaware River Basin studios, McHarg experimented with punch cards (the students rebelled) (Figure 11.5). He was frustrated that the computer at the time could not produce maps with

LANDSCAPE AS INTERACTING PROCESSES

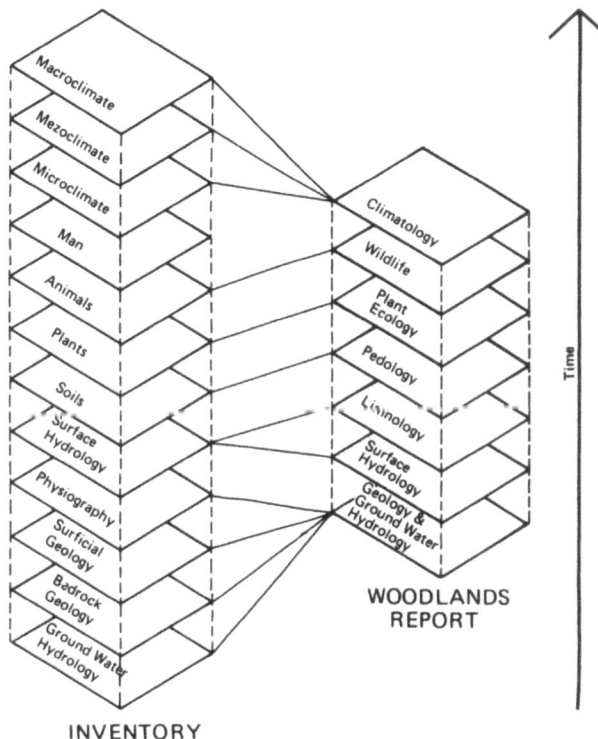

Layer Cake Representation of Phenomena

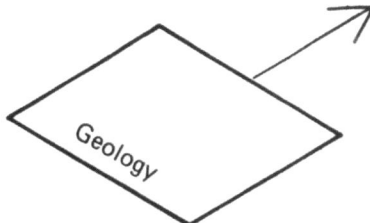

Phenomena Over Time Equals Process

11.1
Ian McHarg's layer-cake diagram, as created by McHarg and Wallace, McHarg, Roberts, and Todd for The Woodlands, Texas, new community in 1974. This diagram illustrates McHarg's approach for collecting and making mapped information about the biophysical and sociocultural processes of a place. He paid special attention to time, suggesting that older elements of the environment occupy the lower layers. Time and ecology are used to organize the data collection and analysis. Ian L. McHarg Collection, The Architectural Archives, University of Pennsylvania.

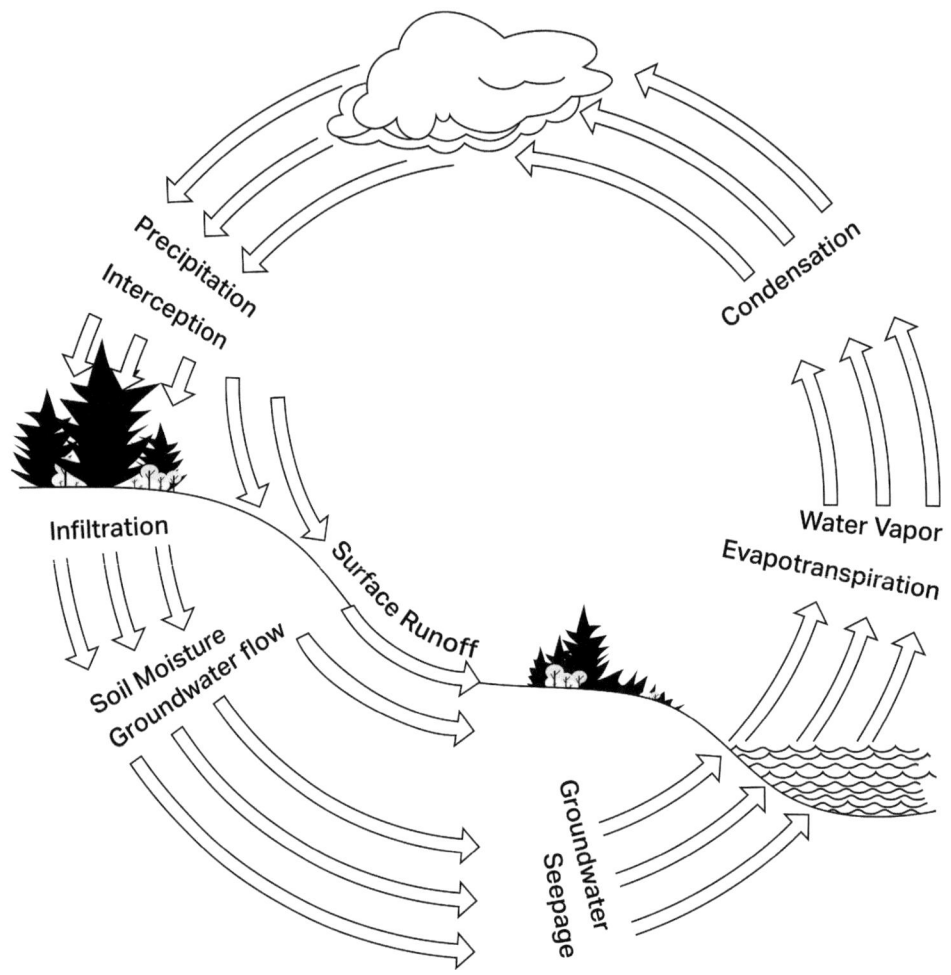

11.2
Each of McHarg's mapped layers represented a slice of time about specific phenomenon based on the best available science at the time. For example, hydrology maps were site-specific representations of the Hydrologic Cycle. The Hydrologic Cycle as diagrammed in Frederick Steiner's The Living Landscape. Recreated by Aaron O'Neill. Credit: Frederick Steiner.

the brilliance of colour that Juneja brought to the map overlays. In the early 1980s, McHarg experimented with Rubylith cut with X-Acto knives that resulted in layers that could be captured by computer (the students rebelled again). Actually, the 'Rubylith' technique presaged scanning technologies. In any case, McHarg eventually adopted Intergraph GIS for his studio teaching in the mid-1980s (Figure 11.6).

Beyond Maps

McHarg's methods for map overlays clearly influenced the advancement of GIS which has become ubiquitous in planning, design, and many areas of research. McHarg supported his approach to ecological design and planning with other representational tools including transects, block diagrams, matrices, and photographs and drawings. In Europe, Scottish polymath Patrick Geddes—a trained biologist and sociologist who became a town planner—developed his valley section, a transect displaying the relationships between elevation and settlements, to help illustrate how the physical world impacts how and where we live. Transects, of course, have a long history. In China, Zhang Zeduan's scroll painting "Along the River during the Qingming Festival" from the Song dynasty displays the daily life of people in the twelfth century. City and country life along the Qingming River are captured in this lively 5.74-yard (5.25-metre) long transect. In the West, early examples of transects

11.3
Delaware River Basin Study II (Fall 1967): Soil and Drainage (F-1). Marker and transfer letters on sepia print, mounted on Chartex. Landscape Architecture and Regional Planning Departmental Records, The Architectural Archives, University of Pennsylvania.

include the German scientist Alexander von Humboldt's drawings of the Andes showing the relationship of vegetation to elevation and English geologist William Smith's strata of Britain. Von Humboldt and Smith sought to display how the surface of the earth reflects forces under our feet. Whereas Von Humboldt travelled the globe, Smith focused on the rocks of Britain. Early in his career, Smith called himself a "mineral surveyor" as a promotion to help wealthy Brits to find coal.

Geddes clearly influenced Tyrwhitt, Benton MacKaye, McHarg, and others.[24] New Urbanists such as Andrés Duany and Elizabeth Plater-Zyberk use transects in their community designs as do many former students of McHarg.[25] As with map overlays, McHarg's contribution to the development of transects as a design and planning tool was his incorporation of ecology to display inter-relationships among organisms and their environments, such as his transects of the New Jerry Shore that he published in *Design with Nature*. In his Penn studios, McHarg had students expand transects through block diagrams, an analysis and representation technique borrowed from geology (Figure 11.7).

McHarg's ecological studies include many diagrams, drawings, and photographs to show relationships and how steps in the process worked. He was especially fond of matrices. For instance, there are seven steps in the suitability analysis process (Table 11.1). McHarg often employed matrices to explain these steps. In a report for Medford Township, New Jersey, Juneja used matrices to illustrate various values and suitability criteria, for instance, for soils (Figure 11.8).[26]

Steps in Suitability Analysis

1 Identify land uses and define the needs for each use.
2 Relate land-use needs to natural factors.

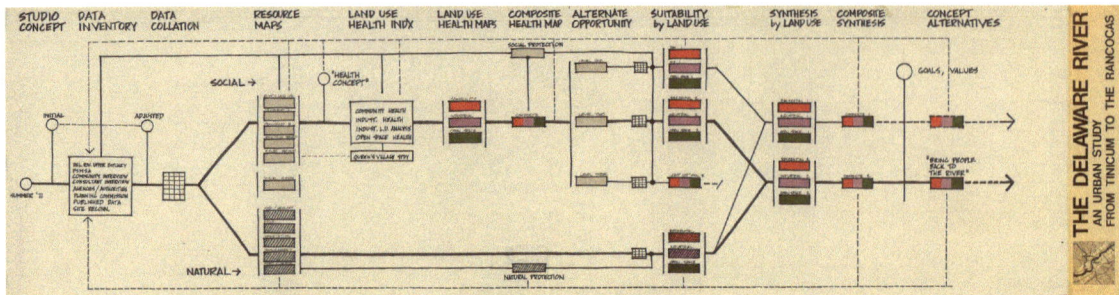

11.4
The production of resource maps was a step in a process to help map health, generate options, and determine suitabilities for various land uses. Landscape Architecture and Regional Planning Departmental Records, The Architectural Archives, University of Pennsylvania.

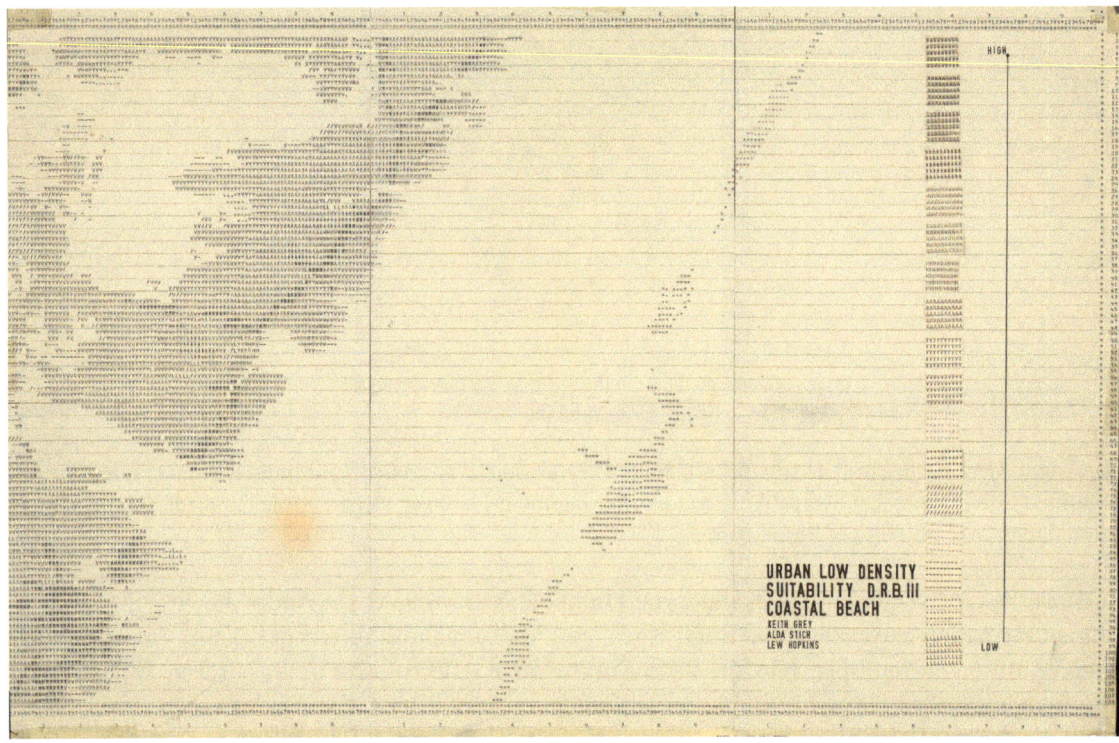

11.5
Suitability map created on an IBM 1401 computer using punch cards for data input. Landscape Architecture and Regional Planning Departmental Records, The Architectural Archives, University of Pennsylvania.

3 Identify the relationship between specific mapped phenomena concerning the biophysical environment and land-use needs.
4 Map the congruences of desired phenomena and formulate rules of combination to express a gradient of suitability. This step should result in maps of land-use opportunities.
5 Identify the constraints between potential land uses and biophysical processes.
6 Overlay maps of constraints and opportunities and, through rules of combination, develop a map of intrinsic suitabilities for various land uses.
7 Develop a composite map of the highest suitabilities of the various land uses.

Source: Adapted from Berger et al. 1977.[27]

Matrices are rather like the box score of a ball game or railroad timetable, allowed for assigning values and cross references. Admittedly, the matrices are not mapping per se, but they were ultimately the

Ian L. McHarg and Mapping Complex Process

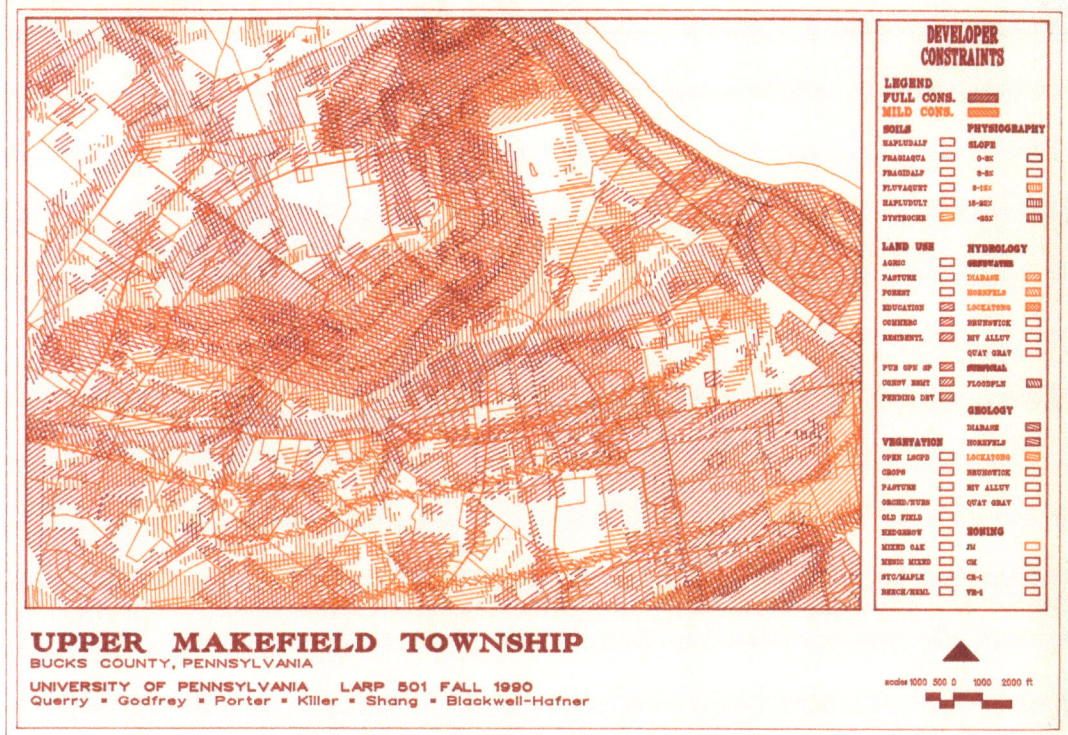

11.6
An example of suitability mapping (or a suitability map) produced using the Integraph GIS system showing Developer Constraints. A companion map showing Developer Opportunities was also produced. Landscape Architecture and Regional Planning Departmental Records, The Architectural Archives, University of Pennsylvania.

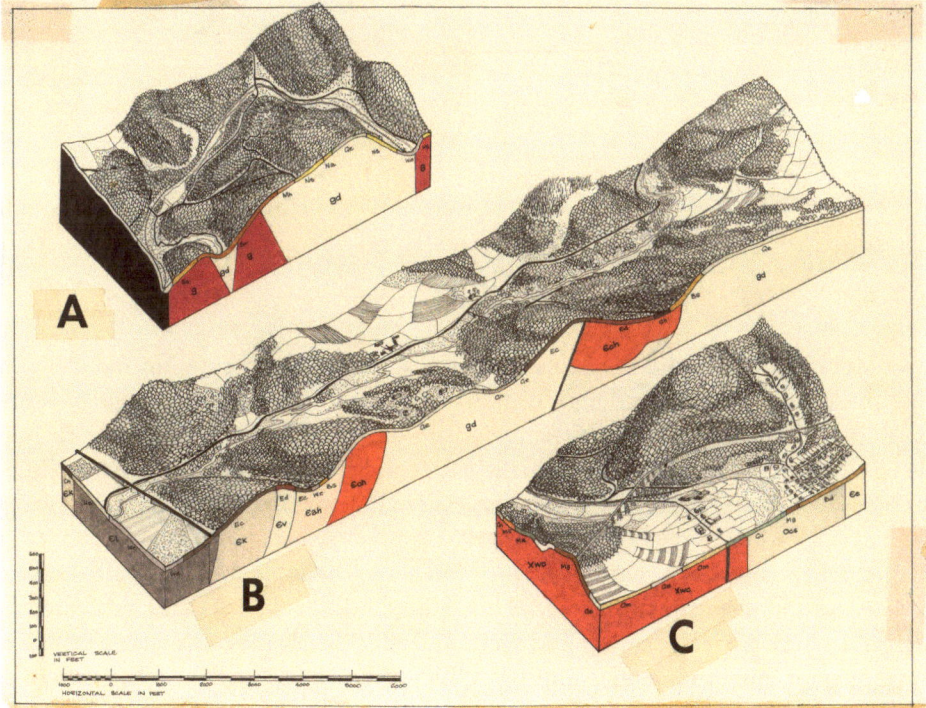

11.7
Block diagram transect from a LARP 501 studio. Ink, marker, pencil, tape, and transfer letters on vellum. Ian L. McHarg Collection of the University of Pennsylvania Architectural Archives.

Table 11.1 Steps in suitability analysis

Identify land uses and define the needs for each use.
Relate land-use needs to natural factors.
Identify the relationship between specific mapped phenomena concerning the biophysical environment land-use needs.
Map the congruences of desired phenomena and formulate rules of combination to express a gradient of suitability. This step should result in maps of land-use opportunities.
Identify the constraints between potential land uses and biophysical processes.
Overlay maps if constraints and opportunities, and though rules of combination develop a map of intrinsic suitabilities for various land use.
Develop a composite map of the highest suitabilities of the various land uses.

Source: Adapted from Berger et al. (1977).

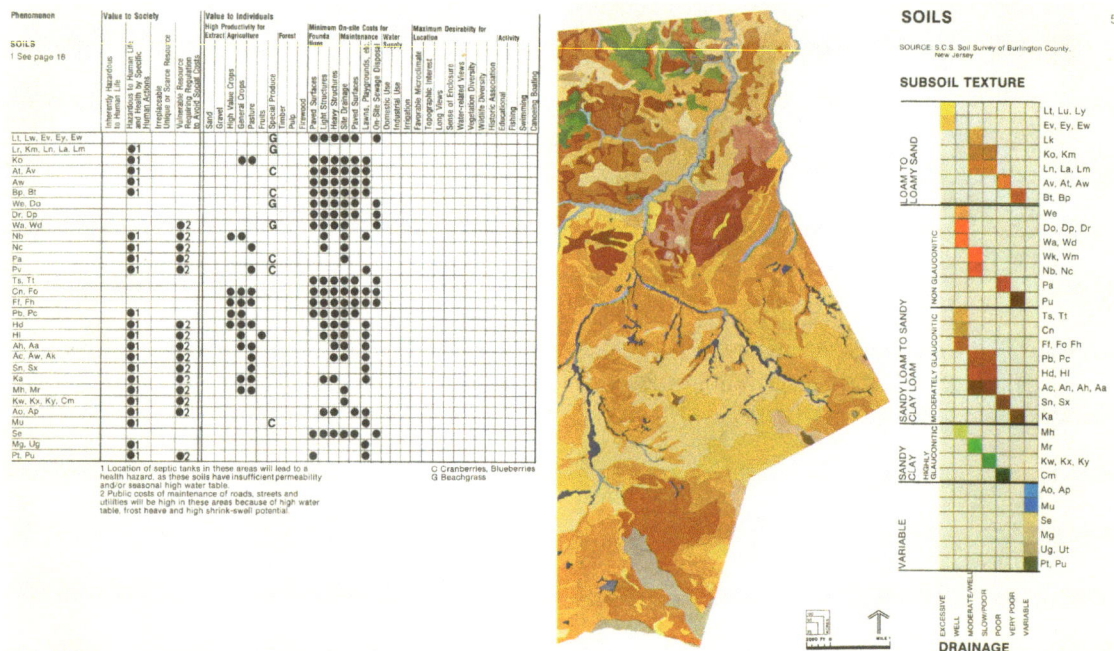

11.8
New Jersey—Medford Township report soils matrix. Narendra Juneja, project director. Ian L. McHarg Collection, The Architectural Archives, University of Pennsylvania.

scorecard for McHarg, Juneja, and their colleagues and students for interpreting and using the maps, and as such were part of the keys to them. Also, this is where some of the more trenchant arguments arose: the values or weight to assign to different factors—say soils, or slope aspect, or economic considerations in terms of suitability—and this is where the awkwardness of social issues came in. For instance, a site might not be fit for solar aspect or less erodible soils, or percolation, but was good for affordability, or mitigation, or development costs.

However, when correlating natural factors and some social factors for health-related issues, which are less squishy in terms of data, McHarg's methods worked well as he demonstrated in the Minneapolis study regarding water and birth rates of 'blue babies' and other deleterious consequences from poor environmental planning. OLIN's recent work in LA with health statistics by census tract when correlated with property values, income, home ownership, and proximity to the LA River and several major freeways is a verification of this—some social data when combined with geo/physical data can be dramatically clear, not debatable, and politically effective (Figure 11.9).[28]

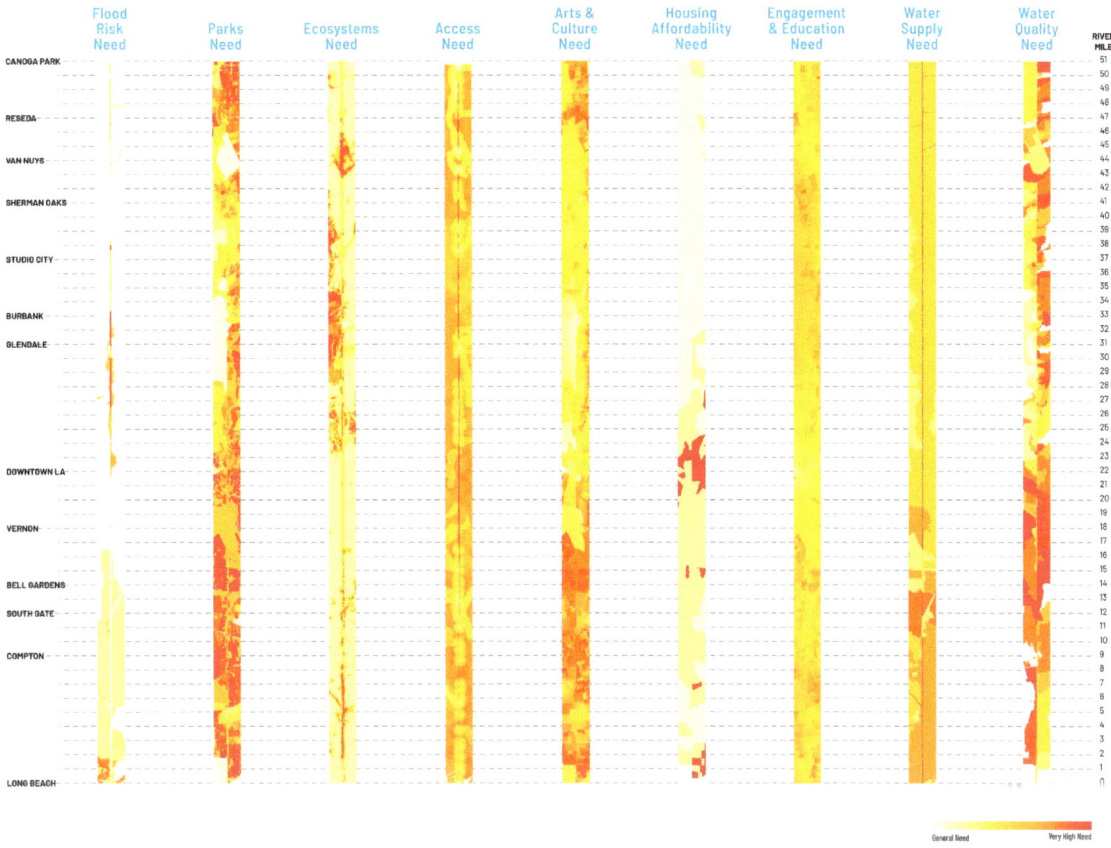

11.9
McHarg's use of mapping and transects remain relevant. For instance, this 2021 study by OLIN illustrates how some areas along the LA River experience unique vulnerabilities, while others contain a variety of assets. Corresponding to the nine master plan goals, these analysis maps explore how a range of needs and opportunities are distributed throughout LA County. Courtesy OLIN with credit to the LA River Master Plan 2021, LA County Public Works, OLIN, Gehry Partners, Geosyntec.

Prospects

Like Steve Jobs who did not invent smartphones or desktop computers but got the whole world to use them, McHarg got the world to use overlay mapping as a way of thinking about data and applying it to design and planning. He was an outspoken advocate for the use of ecological knowledge in design and planning. His ideas remain relevant and necessary.[29] McHarg provided a guide to read landscapes across many scales.

Landscapes are a challenge to represent because they are not static. If architecture is frozen music, as Louis Sullivan declared, then landscape architecture is music thawed and reheated. McHarg provided a score to perform that music with wisdom and creativity.

Notes

1 Manning, W. "The Billerica Town Plan." *Landscape Architecture*, 1913, 3: 108–118; Necktar, L. "Developing Landscape Architecture for the Twentieth Century: The Career of Warren Manning." *Landscape Journal*, 1989, 8(2): 78–91; and Steinitz, C., Parker, P., and Jordon, L. "Hand Drawn Overlays: Their History and Prospective Uses." *Landscape Architecture*, 1976, 9: 444–455.
2 Eliot, C. *Charles Eliot, Landscape Architect*. Boston, MA: Houghton-Mifflin, 1902.
3 Ibid., p. 496.
4 Ibid.
5 Abercrombie, P. and Johnson, T.H. *The Doncaster Planning Scheme*. London, UK: Liverpool University Press, 1922.
6 Regional Planning Staff. *Regional Plan of New York and Its Environs, Vol. 1, The Graphic Regional Plan*. New York, NY: Regional Planning Staff, 1929.

7 Forshaw, J.H. and Abercrombie, P. *County of London Plan*. London, UK: Macmillan, 1943.
8 Steinitz, Parker, and Jordon, *op. cit.*
9 Ibid. Also see Collins, M.G., Steiner, F.R., and Rushman, M. "Land-Use Suitability Analysis in the United States: Historical Development and Promising Technological Achievements." *Environmental Management*, 2001, 28(5): 611–621 and Steiner, F. *The Living Landscape: An Ecological Approach to Landscape Planning*. Washington, DC: Island Press, 2002.
10 Tyrwhitt, J. "Surveys for Planning." In: *Town and Country Textbook*. London, UK: Architectural Press, 1950.
11 Ibid and Steinitz, Parker, and Jordon, *op. cit.*
12 Lyle, J.T. and Stuttgart, F. "Computerized Land Use Suitability Mapping." In: W.J. Ripple (Ed.). *Geographic Information Systems for Resource Management: A Compendium*. Falls Church, VA: American Society for Photogrammetry and Remote Sensing and American Congress on Surveying and Mapping, 1983 and McSheffrey, G. *Planning Derry: Planning and Politics in Northern Ireland*. Liverpool, UK: Liverpool University Press, 1999.
13 Hills, G.A. *The Ecological Basis for Land-Use Planning* (Research Report No. 46). Toronto, Canada: Ontario Department of Lands and Forests, 1961.
14 Belknap, R.K. and Furtado, J.G. *Three Approaches to Environmental Resource Analysis*. Washington, DC: The Conservation Foundation, 1967; Belknap, R.K. and Furtado, J.G. "The Natural Land Unit as a Planning Base: Hills, Lewis, McHarg Methods Compared." *Landscape Architecture*, 1968, 58(2): 145–147; Naveh, Z. and Lieberman, A.S. *Landscape Ecology: Theory and Applications*. 2nd ed. New York, NY: Springer-Verlag, 1994; and Ndubisi, F. "Landscape Ecological Planning." In: G.F. Thompson and F.R. Steiner (Eds.). *Ecological Design and Planning*. New York, NY: John Wiley & Sons, 1997, pp. 9–44.
15 Belknap, R.K. and Furtado, J.G. *Three Approaches to Environmental Resource Analysis. Op. cit.* and Collins, Steiner, and Rushman, *op. cit.*
16 Coombs, D.S. and Thie, J. "The Canadian Land Inventory System." In: M.T. Beatty, G.W. Peterson, and L.R. Swindale (Eds.). *Planning the Uses and Management of the Land*. Madison, WI: American Society of Agronomy, Crop Science Society of America, and Soil Science Society of America, 1979, pp. 909–933.
17 Lewis, P.H. "Quality Corridors in Wisconsin." *Landscape Architecture Quarterly*, 1964, 54(2): 100–107; Lewis, P.H. "The Inland Water Tree." *American Institute of Architects Journal*, 1969, 51(6): 59–63; and Lewis, P.H. *Tomorrow by Design*. New York, NY: John Wiley & Sons, 1996.
18 Ndubisi, F. *Op. cit.*, p. 21.
19 Belknap, R.K. and Furtado, J.G. *Three Approaches to Environmental Resource Analysis. Op. Cit.*; Belknap, R.K. and Furtado, J.G. "The Natural Land Unit as a Planning Base: Hills, Lewis, McHarg Methods Compared." *Op. Cit.*; and Steinitz, Parker, and Jordon, *op. cit.*
20 Collins, Steiner, and Rushman, *op. cit.*
21 Belknap, R.K. and Furtado, J.G. *Three Approaches to Environmental Resource Analysis. Op. cit.*; Belknap, R.K. and Furtado, J.G. "The Natural Land Unit as a Planning Base: Hills, Lewis, McHarg Methods Compared." *Op. cit.*; Gordon, S.I. *Computer Models in Environmental Planning*. New York, NY: Van Nostrand Reinhold, 1985; McHarg, I.L. *Design with Nature*. Garden City, NY: Natural History Press, 1969; and McHarg, I.L. and Steiner, F.R. (eds.). *To Heal the Earth: Selected Writings of Ian L. McHarg*. Washington, DC: Island Press, 1998.
22 McHarg, I.L. *Design with Nature. Op. cit.*
23 McHarg, I.L. *A Quest for Life*. New York, NY: John Wiley & Sons, 1997, p. 321.
24 Steiner, F. and McSherry, L. "Observation, Reflection, Action." *Landscape and Urban Planning*, 2017, 166: 55–56.
25 Duany, A. and Talen, E. "Transect Planning." *Journal of the American Planning Association*, 2002, 68(3): 245–266; Han, S.M. "The Use of Transects for Resilient Design: Core Theories and Contemporary Projects." *Landscape Ecology*, 2021, 36(6): 1–16; and Weller, R. and Talarowski, M. *Transects: 100 Years of Landscape Architecture and Regional Planning at the School of Design of the University of Pennsylvania*. San Francisco, CA: Applied Research + Design, 2014.
26 Juneja, N. *Medford: Performance Requirements for the Maintenance of Social Values Represented by the Natural Environment of Medford Township, New Jersey*. Philadelphia: Center for Ecological Research in Planning and Design, University of Pennsylvania, 1974.
27 Berger, J., Johnson, A., Rose, D., and Skaller, P. *Regional Planning Notebook* (LARP 501 Course Guidelines). Philadelphia: Department of Landscape Architecture and Regional Planning, University of Pennsylvania.
28 Steiner, F., Weller, R., M'Closkey, K., and Fleming, B. (eds.). *Design with Nature Now*. Cambridge, MA: Lincoln Institute of Land Policy, 2019.
29 Cohen, W. *Ecohumanism and the Ecological Culture: The Educational Legacy of Lewis Mumford and Ian McHarg*. Philadelphia, PA: Temple University Press, 2019; John-Alder, K. *Ian McHarg and the Search for Ideal Order*. London, UK and New York, NY: Routledge, 2019; Steiner, Weller, M'Closkey, and Fleming (Eds.). *Op. cit.*;

Steiner, F. and Fleming, B. "Design with Nature at 50: It's Enduring Significance to Socio-Ecological Practice and Research in the Twenty-First Century." *Socio-Ecological Practice Research*, 2019, 1: 173–177; and Yang, B. *Landscape Performance: Ian McHarg's Ecological Planning in The Woodlands, Texas*. London, UK and New York, NY: Routledge, 2019.

References

Abercrombie, P. and Johnson, T.H. *The Doncaster Planning Scheme*. London, UK: Liverpool University Press, 1922.

Belknap, R.K. and Furtado, J.G. "The Natural Land Unit as a Planning Base: Hills, Lewis, McHarg Methods Compared." *Landscape Architecture*, 58(2) (1968): 145–147.

Belknap, R.K. and Furtado, J.G. *Three Approaches to Environmental Resource Analysis*. Washington, DC: The Conservation Foundation, 1967.

Berger, J., Johnson, A., Rose, D., and Skaller, P. *Regional Planning Notebook* (LARP 501 Course Guidelines). Philadelphia, PA: Department of Landscape Architecture and Regional Planning, University of Pennsylvania.

Cohen, W. *Ecohumanism and the Ecological Culture*: *The Educational Legacy of Lewis Mumford and Ian McHarg*. Philadelphia, PA: Temple University Press, 2019.

Collins, M.G., Steiner, F.R., and Rushman, M. "Land-Use Suitability Analysis in the United States: Historical Development and Promising Technological Achievements." *Environmental Management*, 28(5) (2001): 611–621.

Coombs, D.S. and Thie, J.. "The Canadian Land Inventory System." In: M.T. Beatty, G.W. Peterson, and L.R. Swindale (Eds.). *Planning the Uses and Management of the Land*. Madison, WI: American Society of Agronomy, Crop Science Society of America, and Soil Science Society of America, 1979, pp. 909–933.

Duany, A. and Talen, E. "Transect Planning." *Journal of the American Planning Association*, 68(3) (2002): 245–266.

Eliot, C. *Charles Eliot, Landscape Architect*. Boston, MA: Houghton-Mifflin, 1902.

Forshaw, J.H. and Abercrombie, P. *County of London Plan*. London, UK: Macmillan, 1943.

Gordon, S.I. *Computer Models in Environmental Planning*. New York, NY: Van Nostrand Reinhold, 1985.

Han, S.M. "The Use of Transects for Resilient Design: Core Theories and Contemporary Projects." *Landscape Ecology*, 36(6) (2021), 1–16.

Hills, G.A. *The Ecological Basis for Land-Use Planning* (Research Report No. 46). Toronto: Ontario Department of Lands and Forests, 1961.

John-Alder, K. *Ian McHarg and the Search for Ideal Order*. London, UK and New York, NY: Routledge, 2019.

Juneja, N. *Medford*: *Performance Requirements for the Maintenance of Social Values Represented by the Natural Environment of Medford Township, New Jersey*. Philadelphia: Center for Ecological Research in Planning and Design, University of Pennsylvania, 1974.

Lewis, P.H. "Quality Corridors in Wisconsin." *Landscape Architecture Quarterly*, 54(2) (1964): 100–107.

Lewis, P.H. "The Inland Water Tree." *American Institute of Architects Journal*, 51(6) (1969): 59–63.

Lewis, P.H. *Tomorrow by Design*. New York, NY: John Wiley & Sons, 1996.

Lyle, J.T. and Stuttgart, F. "Computerized Land Use Suitability Mapping." In: W.J. Ripple (Ed.). *Geographic Information Systems for Resource Management*: *A Compendium*. Falls Church, VA: American Society for Photogrammetry and Remote Sensing and American Congress on Surveying and Mapping, 1983.

Manning, W. "The Billerica Town Plan." *Landscape Architecture*, 3 (1913): 108–118.

McHarg, I.L. *A Quest for Life*. New York, NY: John Wiley & Sons, 1997.

McHarg, I.L. *Design with Nature*. Garden City, NY: Natural History Press, 1969.

McHarg, I.L. and Steiner, F.R. (eds.). *To Heal the Earth*: *Selected Writings of Ian L. McHarg*. Washington, DC: Island Press, 1998.

McSheffrey, G. *Planning Derry*: *Planning and Politics in Northern Ireland*. Liverpool, UK: Liverpool University Press, 1999.

Naveh, Z. and Lieberman, A.S. *Landscape Ecology*: *Theory and Applications*. 2nd ed. New York, NY: Springer-Verlag, 1994.

Ndubisi, F. "Landscape Ecological Planning." In: G.F. Thompson and F.R. Steiner (Eds.) *Ecological Design and Planning*. New York, NY: John Wiley & Sons, 1997, pp. 9–44.

Necktar, L. "Developing Landscape Architecture for the Twentieth Century: The Career of Warren Manning." *Landscape Journal*, 8(2) (1989): 78–91.

Regional Planning Staff. *Regional Plan of New York and its Environs, Vol. 1, The Graphic Regional Plan*. New York, NY: Regional Planning Staff, 1929.

Steiner, F. *The Living Landscape*: *An Ecological Approach to Landscape Planning*. Washington, DC: Island Press, 2002.

Steiner, F. and Fleming, B. "Design with Nature at 50: It's Enduring Significance to Socio-Ecological Practice and Research in the Twenty-First Century." *Socio-Ecological Practice Research*, 1 (2019): 173–177.

Steiner, F. and McSherry, L. "Observation, Reflection, Action." *Landscape and Urban Planning*, 166 (2017): 55–56.

Steiner, F., Weller, R., M'Closkey, K., and Fleming, B. (eds.). *Design with Nature Now*. Cambridge, MA: Lincoln Institute of Land Policy, 2019.

Steinitz, C., Parker, P., and Jordon, L. "Hand Drawn Overlays: Their History and Prospective Uses." *Landscape Architecture*, 9 (1976): 444–455.

Tyrwhitt, J. "Surveys for Planning." In: *Town and Country Textbook*. London, UK: Architectural Press, 1950.

Weller, R. and Talarowski, M. *Transects: 100 Years of Landscape Architecture and Regional Planning at the School of Design of the University of Pennsylvania*. San Francisco, CA: Applied Research + Design, 2014.

Yang, B. *Landscape Performance: Ian McHarg's Ecological Planning in The Woodlands, Texas*. London, UK and New York, NY: Routledge, 2019.

12 Drawing Experiments for Representing Landscape

Noémie Lafaurie-Debany, Javier González-Campaña, and Balmori Associates

BAL/LAB

Since 2006, Balmori Associates, Landscape and Urban Design, based in New York City, has been divided into two parts. The first is a landscape practice that investigates landscape as a constructed space. The second part, BAL/LAB, is a collection of research and experiments: collaborations across disciplines, exploration of new technologies, projects invented by us, temporary projects, floating landscapes, and the zero-waste city are some of the BAL/LABs. Another one deals with the challenges of representing landscape.

In landscape architecture, representation has become the subject of contention and much discussion. While 3D modelling and rendering software have been a catalyst for change across the field of design, nowhere have the conventions of representation been called into question more than in landscape architecture, which is undergoing a process of reinvention. With the onset of rapid urbanization and our shifting relationship with nature, landscape architecture has proved a potent lens for expressing a wider dialogue taking place in the world. It is, however, only through the introduction of innovative forms of representation—whether digital, analogue, or hybrid—that one most vividly sees the emergence of the new.

In Representation BAL/LAB or during the design of projects, we are developing drawing experiments for rendering landscape. We are researching ways one apprehends the landscapes we are designing. Our intention is to make the viewer more conscious of the space we are creating rather than the objects that are in the space. To this end, we seek to erase the edges between objects to focus on what is in between, their interface. One method we have developed is using a dot matrix inspired by the halftone and the Ben Day process as used by Roy Lichtenstein. Another method focuses on patterns rather than contours, like what one can find in twentieth-century French artist Pierre Bonnard's paintings. In our drawings, we seek to represent the spatial quality of our designs as well as the character and atmosphere of the landscapes.

The main tenet of our work is to set up a different type of relationship between ourselves and each of the elements of nature: soil, water, air, plants, and animals. We want to change our ways of dealing with them, treat them as parts of ourselves. This principle is expressed in drawings, for example, with the soil and roots of plants drawn with the same importance given to the visible canopy of the trees. Illustrating how people could use the space is represented with silhouettes—transparent black or white figures—rather than realistic photos of people. This is to reduce bias, allow the viewer of the image to imagine herself in the landscape, and focus on the space rather than the latest fashion. Some perspectives, and by extension projects and ideas, become dated by the people that inhabit them and what they wear. We have also tested representing activities with hand-drawn figures to stress the flexible use of the space and the temporary nature of the activities.

Process

Drawing is the language of ideas. It is how designers generate and test ideas and communicate the intentions of the design to the team, the public, and the client. Ideas may materialize in plan or section

DOI: 10.4324/9781003183402-12

view, a combination of the two, or a perspective. They may be produced by sketching with pencils on trace paper, or while transforming geometries in Rhino or other 3D software, or while doing a collage or a physical model.

In our practice, analogue and digital processes are intertwined throughout the design of the project. The evolution of the design does not follow a linear process that would start with rougher, less-defined, intention-driven lines to more detailed and precise computer-generated ones. Instead, representation—in a diverse range of techniques—is integral and fully integrated into the design process. Perspectives or computer-generated renders are produced in-house through the development of the design. Some design practices may hire renderers at the end of the design process to represent the project, and it often produces hyper-realistic images, detached from the project's concept. Development of rendering software like Lumion has allowed perspective images to be easily produced at any stage of the design. In our practice, we have used Lumion and then further altered the images created with the software. Recently, we have worked with a visualization company that integrates the design team early in the design process, testing ideas and rendering alternatives.

During the COVID-19 pandemic, our studio had to rely primarily on digital tools to communicate and design ideas: meetings in the conference room were replaced by Microsoft Teams meetings, the pin-up boards by Miro "online whiteboards," and live redlining during design review happened on the screen and became ephemeral.

While preparing Balmori Associates' records to be sent to Yale University's Sterling Memorial Library, we prioritized the archiving of process images, not just the final deliverables. As expected, through the 30 years of production of ideas since Balmori Associates was founded by Diana Balmori, the balance between analogue and digital shifted from a majority of paper and physical models in the early years to mostly digital. One of the challenges of capturing and recording ideas and the process, and not just final deliverables, is that they may reside in versions of a file generated with a licensed software that only owners of such software can access. In that context, the redlining or sketches over printed material has been crucial in offering a more universally accessible outlook on the design process.

Time

A landscape is never the same twice. It changes through the seasons. It changes through the years, and it is constantly changing with the play of shadow, the tides, and the clouds. But landscape renderings often fix the landscape in time: one hour of the day, one season, and a certain number of years after the landscape would be first planted. For projects that will be implemented in phases, it is custom to see a series of drawings representing the landscape—year 0, year 3, year 5, year 10, etc. Sometimes, that time sequence is driven by construction management requirement or is part of the ecological succession, like when restoring a natural ecosystem. "This lack of fixity is landscape asset," wrote Diana Balmori in her *Landscape Manifesto*.[1] Many of the drawing experiments have revolved around the representation of time in landscape.

In summer 2021, Balmori Associates is opening a garden installation at the Jardins de Métis International Garden Festival called "Choose Your Own Adventure." Rethinking our connection to nature, after living in lockdown during the COVID-19 pandemic, can start with appreciating natural phenomena: the gusty wind, the wet bark, the musky shade, the dry air, the sweet smell, the hot stone, the crunchy gravel, etc. The project evokes the ever-changing quality of the landscape and seeks to create a space that can only be experienced and not represented or photographed. The garden challenges the still frame images posted on the worldwide web reminiscent of the eighteenth-century picturesque. Malcolm Andrews described tourists seeking the ideal landscapes as "'fixing' them as pictorial trophies in order to sell them or hang them up in frames on their drawing room walls"[2] (aka Instagram of the twenty-first century!). But a landscape never happens twice, and its lack of fixity and hyper-sensorial experiences are heightened through a simple matrix that forms the garden. The representation of the garden in the competition submission remained diagrammatic on purpose: running East/West bands of planting intercepts, North/South bands of different hard materials. The garden calls the visitor to choose their own adventure, smell, touch, listen, taste, and see.

Representing a landscape that is constantly changing is certainly challenging. We have tested animation, but find ourselves limited by the specific tech required to produce an animation that carries our ideas. We have produced sections and diagrams to address the issue of a phased landscape and the growth of the plants through time. We have represented the same view at different seasons or the same view divided into four seasons.

Frame and Peripheral Vision

Diana Balmori wrote in *Drawing and Reinventing Landscape* that "Landscape architecture is an art of peripheral vision. Peripheral vision is essential for understanding and appreciating landscape; central vision alone cannot capture it."[3] To explore this, vision scientist Denis Pelli and Balmori Associates' staff set up an experiment to measure how restricting the observer's field of view affects the observer's experience of the beauty of a landscape. The viewing devices chosen for this demonstration are a tube and truncated cone (with both ends cut off). The results show that restricting the observer's peripheral vision reduced the viewing pleasure.

In 2011, Balmori Associates implemented the viewing cone described above as a series of planes with a circular opening, the void gradually rising from the ground at the Jardins de Métis Garden Festival. When progressing through the frames towards the St. Lawrence River, focusing on the floating islands, the field of view opens, the horizon gets wider, and infinite space offers itself to the viewer.

After testing the role of peripheral vision by experiencing an existing landscape and the design of an installation that demonstrates the importance of peripheral vision in the pleasure given by a landscape, we pursued to represent landscape with peripheral vision. An initial series of tests that consisted of applying a homogeneous filter to the periphery of the image failed. We have yet to succeed in doing this, and the deformation of the objects in the image was more successful.

The frame is an important element in the perceptual experience of landscape, and the viewport in 3D modelling software plays a similar role in the representation of landscape. By creating an outline, the frame defines a field and creates a view. "Looking at a view"—'what does that mean?' asks W. J. T. Mitchell in his essay called "Landscape and Invisibility."[4] The view is the combination of three elements: the collection of individual objects in the visual field—a tree, a flower, a leaf, or a bench—the relations between them, the connections they create in space, and the imagination of the observers, their recollections, and knowledge. Refer to Figures 12.1–12.9, illustrating a variety of Balmori Associates' drawing styles.

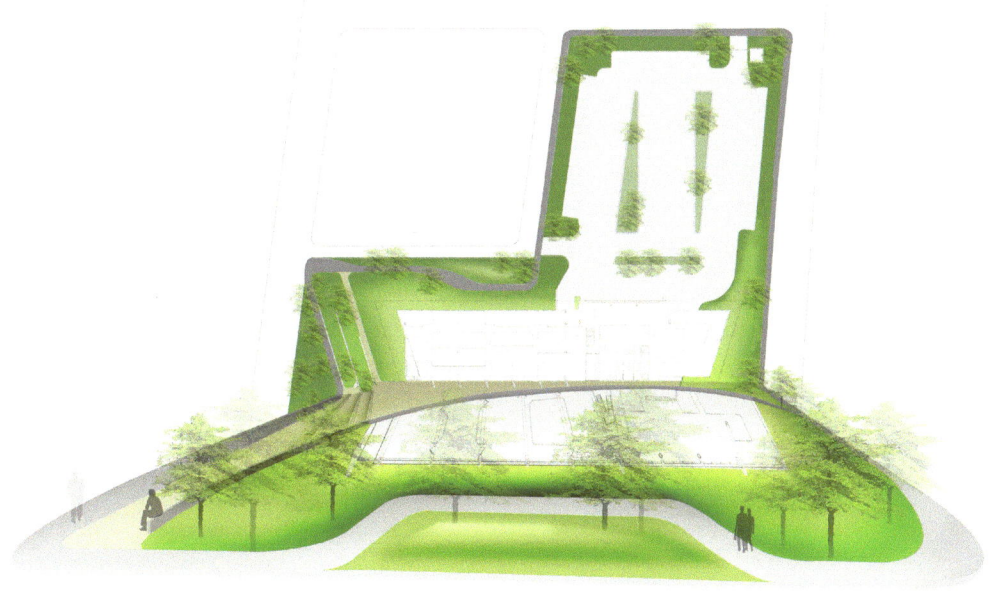

12.1
Harrisburg. Drawing combining plan and perspectives for the Courthouse, Harrisburg, PA.

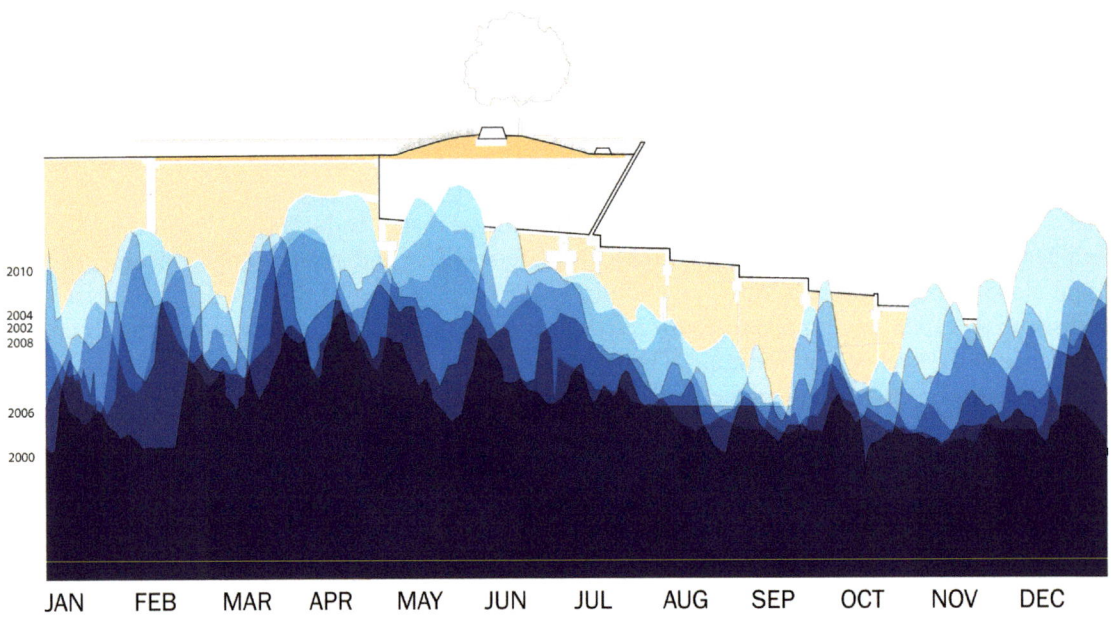

12.2
Bealestreet Flood. Section of Beale Street Landing in Memphis, TN, in relation to Mississippi water levels.

12.3
Binney3. Four seasons rendered across the view of a residential project in Greenwich, CT.

Drawing Experiments for Landscape

12.4
BRIT. Dot matrix applied to the entrance of the Botanical Research Institute of Texas, TX.

12.5
Madrid Nuevo Norte, Perspective Legendre. Digital and analogue techniques illustrate the space and the activities taking place in Calle de Mauricio Legendre in Madrid Nuevo Norte, Spain.

12.6
Madrid Nuevo Norte Section. Section displaying prominently the roots of the trees in Madrid Nuevo Norte, Spain.

12.7
Métis CYOA Time-Process. Diagram illustrating three stages of the "Choose Your Own Adventure" garden at Jardins de Métis International Garden Festival, when first planted and how it gets modified with people's movement in it.

Drawing Experiments for Landscape

12.8
Soundwaves Rollview. The scroll captures the experience of walking through the garden for Beijing Garden Expo.

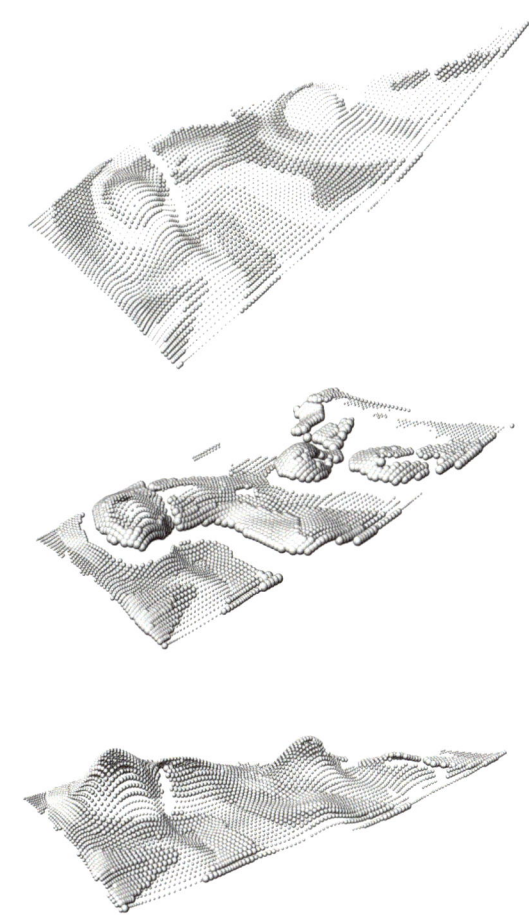

12.9
Soundwaves Growth. The diagrams show the growth of the plants in relation to site conditions—slope and sun exposure.

Notes

1 Balmori, D. *A Landscape Manifesto*. New Haven, CT: Yale University Press, 2012, p. 225.
2 Andrews, M. *The Search for the Picturesque*. Stanford, CA: Stanford University Press, 1989, p. 67.
3 Balmori, D. *Drawing and Reinventing Landscape*. Chichester, UK: John Wiley & Sons Ltd., 2014, p. 76.
4 Mitchell, W.J.T. "Landscape and Invisibility: Gilo's Wall and Christo's Gates." In: D. Harris and D. Fairchild Ruggles (Eds.). *Sites Unseen: Landscape and Vision*. Pittsburgh, PA: University of Pittsburgh Press, 2007, pp. 33–44.

References

Andrews, M. *The Search for the Picturesque*. Stanford, CA: Stanford University Press, 1989.
Balmori, D. *A Landscape Manifesto*. New Haven, CT: Yale University Press, 2012.
Balmori, D. *Drawing and Reinventing Landscape*. Chichester, UK: John Wiley & Sons Ltd., 2014.
Mitchell, W.J.T. "Landscape and Invisibility: Gilo's Wall and Christo's Gates." In: D. Harris and D. Fairchild Ruggles (Eds.). *Sites Unseen: Landscape and Vision*. Pittsburgh, PA: University of Pittsburgh Press, 2007, pp. 33–44.

13 Peter Walker
The Growth of Representation

Peter Walker

Over many years, our office has explored multiple forms of representation. By 1950, modern American offices had developed two main categories of graphics: design representation and working drawings. Although working drawings and specifications that direct bidding and legal representation modernized with relative speed, the effort to first communicate the design and, particularly, the intended design character has seen a slower evolution. It is these discoveries and experiments that are the subject of this chapter.

I believe landscape architecture is an art, but unlike painting or sculpture, it is an art that must be graded, built, and planted in order to be realized. So how have we communicated intent to others and to ourselves? Looking back carefully over more than 60 years of practice has shown me the ways we have progressed and the events, discoveries, and techniques that have brought us through a long road of invention to this point in time.

Since the beginning of my practice in 1955, landscape architects have been trying to find ways to represent landscape architecture designs in order to more effectively communicate with clients, architects, and public bodies. Architecture had for years developed ways of drawing in plans, elevations, sections, axons, models, and perspectives, while landscape design depended primarily on plans or maps and an occasional section. And yet landscape architecture is an art of milieu, in which colour, form, and, particularly, space are dependent on seasonal growth and changes of light and shadow. These are its most important aspects, none of which can be effectively mapped. Stanley White said early on that the landscape architect must be able to imagine designed space from the most intimate garden to the most extended landscape.[1] Only much later did I realize the complex difficulty of representing this range of projects.

Early Landscape Drawings

At the time of my graduation, the most famous landscape architects only added axonometric (Eckbo, Royston, and Williams) and cartoon sketches (Church and Halprin) to their traditional plans. Even simple cross-sections were a rarity. The first professional drawings that I made were in 1954. I was then working for Lawrence Halprin, and so understandably, these were in the style of the Halprin office and reflected Halprin's work at the Thomas Church office, where he had entered practice in the late 1940s (see Figure 13.1).

"Corporate" Drawings

In 1956, while working in the Sasaki office in Watertown, Massachusetts, we made drawings for the Upjohn Headquarters in Kalamazoo, Michigan, which was influenced by the design of the Noguchi gardens at Connecticut General Headquarters in Hartford, Connecticut, a site I had seen on a field trip while a student at the Harvard Graduate School of Design. At the Sasaki office, first in 1956 and then after our graduation, we gradually developed a "district" style as we worked on universities, corporate campuses, and urban renewal. Note the radical use of press on 'Zip-A-Tone'[2] to represent the ground plane. This look was known as "corporate style" from the preliminary drawings for such

DOI: 10.4324/9781003183402-13

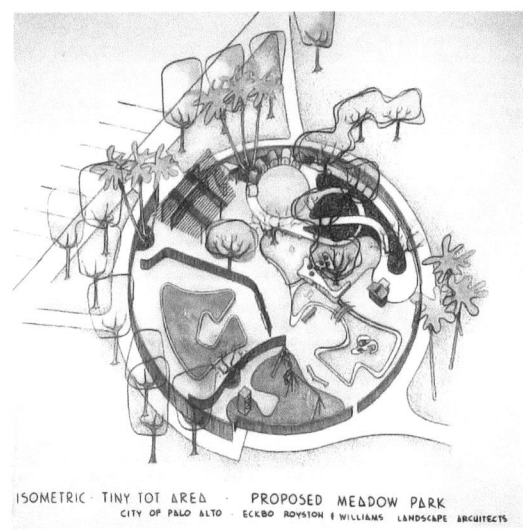

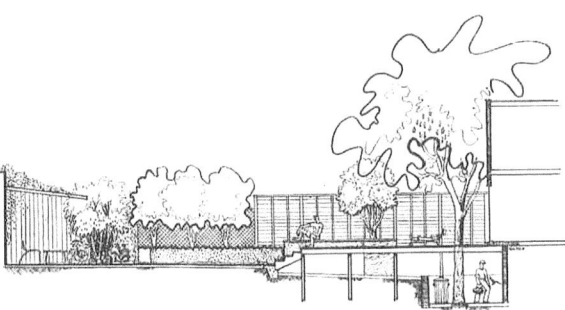

13.1
Collection of Early Works. The first professional drawings that I made were in the style of the Halprin office. At the time, the most famous landscape architects only added axonometric and cartoon sketches to their traditional plans. Even simple cross-sections were a rarity.

projects as the Upjohn Headquarters, Kalamazoo, Michigan, 1957–1961, and, Foothill College, Los Altos Hills, California, 1957–1960 (see Figure 13.2).

Exceptional Drawings

Over the first decade of the Sasaki Walker and Associates office in Sausalito, California (1960–1970), we often collaborated with the remarkable landscape architect Bill Johnson,[3] who helped us make plans and perspective drawings for all of our most important work, including competitions. Bill had been doing these drawings since we were classmates at Harvard. Bill's perspective drawings at all scales have influenced the drawing of much of SWA's work to this day, and he is still one-of-a-kind (see Figure 13.3).

Artist Drawing for Character

From 1996 to 2015, we worked with artist Chris Grubbs, whom we met through the architects Skidmore Owings & Merrill. His drawings had been used in architectural and urban design projects, but not so much in landscape. Nevertheless, Chris had learned to go beyond drawing buildings to include their settings. Often to deepen our presentations, Chris could focus on landscape in order to express landscape character. In the period when we were beginning to get institutional projects, like the National September 11 Memorial, character became a major issue (see Figure 13.4).

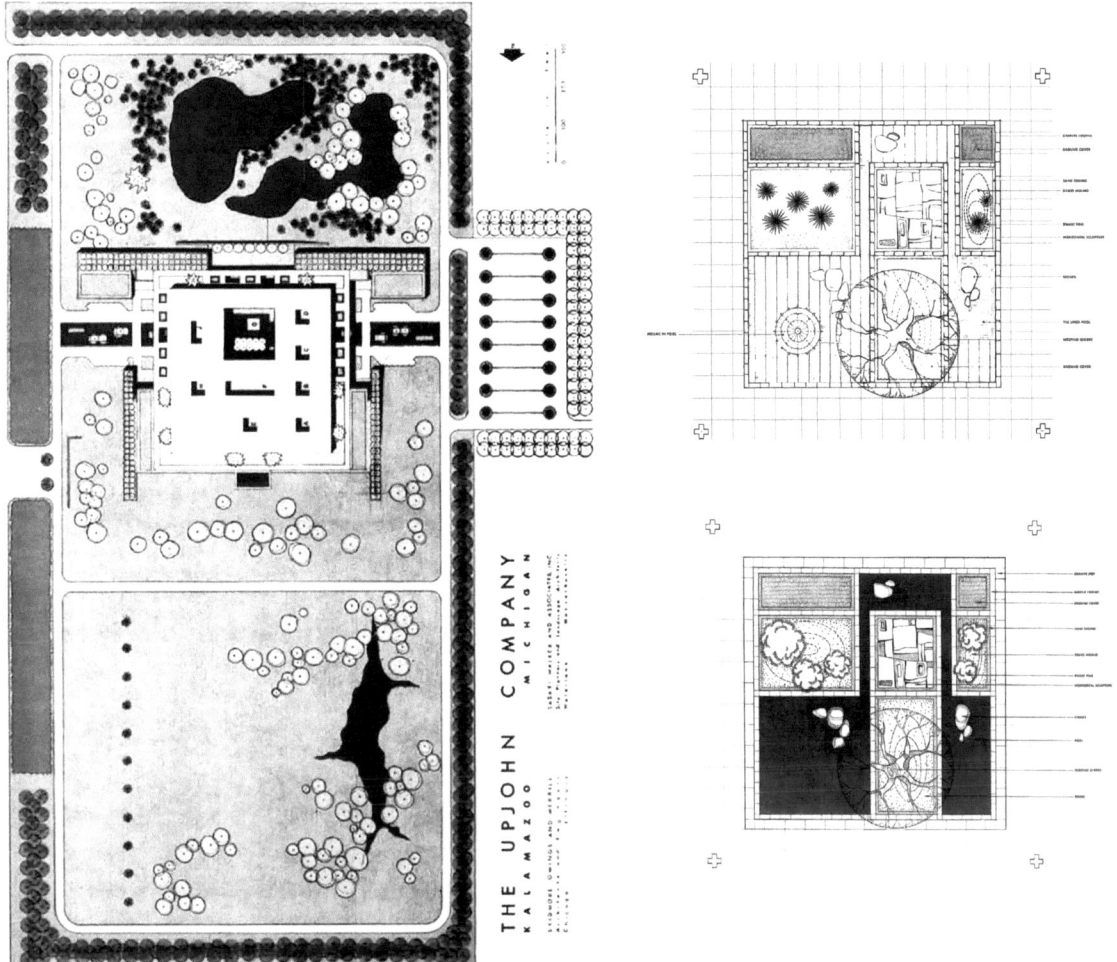

13.2
Collection of "Corporate" Drawings. At the Sasaki office, we developed a "district" style as we worked on universities, corporate campuses, and urban renewal. This look was known as "corporate style" from the preliminary drawings for such projects as the Upjohn Headquarters, Kalamazoo, Michigan.

Photography to Illustrate Character, Scale, and Plant Materials

Another communication technique that we explored was the use of photographic slides to augment and bring character to our presentation plans and sections. In this pursuit, we collected tens of thousands of slides filed in plastic sheets. At first, these photographs were taken by our partners and staff in their travels over the U.S. and abroad. The categories included contemporary and classical gardens and landscape materials of all kinds such as paving, walls, and plant composition. At some point, we employed a professional staff photographer, Jerry Campbell, to expand our slide collection and to document construction progress, which was, by then, becoming complicated work. Through this technique, we began to make slide shows combining both drawings and photos (now called Power-Points) for our conversations with clients. Interestingly, these images expanded to inspire our design work. We have also used photography to record models (see Figure 13.5).

Models Empower the Curriculum

When I became chair at Harvard in 1978, the faculty decided to create an additional one-year curriculum for graduate students without a previous design degree. These new students were mostly more broadly educated than students with Bachelors in Landscape Architecture, but they had not learned

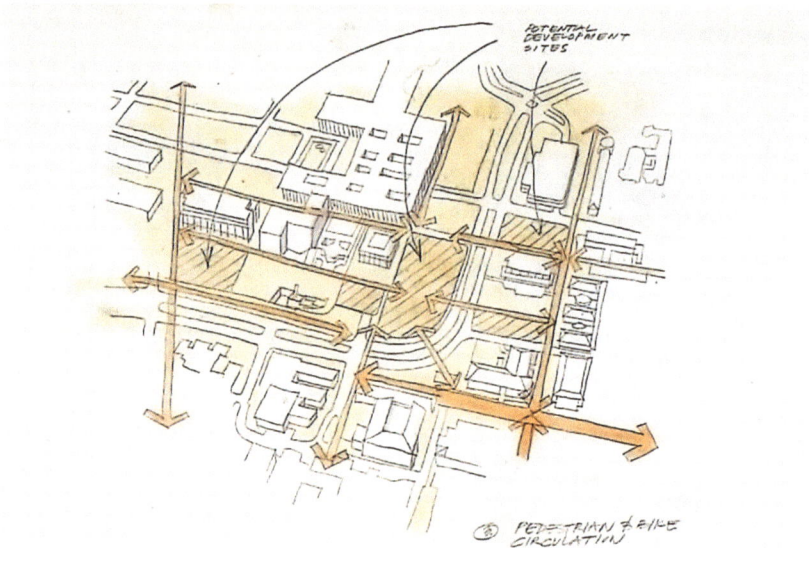

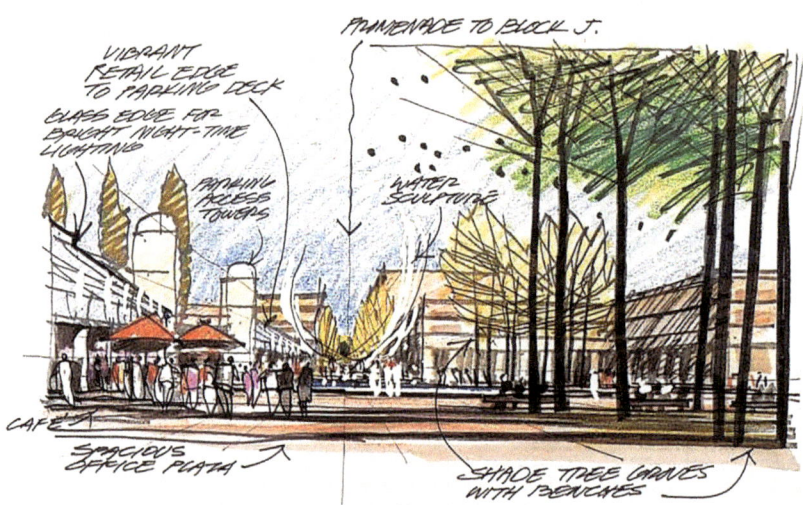

13.3
Collection of Exceptional Drawings. Over the first decade of the Sasaki Walker and Associates, we often collaborated with the remarkable landscape architect Bill Johnson, who helped us make plans and perspective drawings for all of our most important work, including competitions.

13.4
Artist Drawing for Character. In the period when we were beginning to get institutional projects, character became a major issue. From 1996 to 2015, we worked with artist Chris Grubbs. Often, to deepen our presentations, Chris could focus on landscape in order to express landscape character.

13.5
Collection of Illustrative Photos. We explored the use of photographic slides to augment and bring character to our presentation plans and sections. We began to make slide shows combining both drawings and photos for our conversations with clients to illustrate character, scale, and plant materials.

13.6
Detail Model. We were, perhaps, the only landscape office to use models almost exclusively to represent our work. These models were augmented by full-scale mock-ups and details created in our parking lot, a form of large model-making.

how to draw. We tried several remedies, for example, enrolling them in (Harvard's art school) mechanical drawing classes in the summer and life drawing classes in the evenings. None of these remedies worked. The students who could still not compete with the BLA students in drawing (even in plan) were becoming depressed, some even dropping out of our program. It was then that we decided on making detailed models with in-scale trees, careful representation of grade, colour, and realistic development of the ground plane. These models could be observed as "real"—although miniature—three-dimensional spaces.

At that time, the BLA students were mostly still making plan drawings and were not adept at model-making. Still, their interest was strong, and they quickly became ace model-makers.

A number of students from both groups joined our firm (PWP) after their graduation. During this period, we were, perhaps, the only landscape office to use models almost exclusively to represent our work. These models were augmented by full-scale mock-ups and details created in our parking lot, a form of large model-making (see Figure 13.6).

Introduction of Computers

It is the utilization of computers that has had the largest impact on our ability to represent the intent of our design. Until David Walker introduced a primitive form of Vectorworks® into the office in 1992–1994, we were drawing everything by hand and making models. Our then-client Steve Jobs (Pixar) asked us if designers would welcome an alternate to AutoCAD® which, at that time, was exclusively

13.7
Introduction of Computers. The ability to draw on the computer gradually took over our communication methods including construction drawings. These drawings were not meant to be realistic, but they were exact and clear and quick.

required by the engineers. AutoCAD was only available on PCs and offered no colour or graphic nuance. We then switched to using the Mac entirely, but we had no printer until Hewitt Packard introduced one in about 1995. It cost $10,000, a large sum in those days. From that point, Photoshop and MiniCAD (by Adobe) added colour, shading, and flexible hue.

The ability to draw on the computer gradually took over our communication methods, including construction drawings. These drawings were not meant to be realistic, but they were exact and clear and quick, and soon took over our hand-drawing needs, helped with scaling and copying, and greatly reduced the making of models. All of this occurred before the introduction of the Internet. By 1996, few of us utilized hand-drawings at all. Ironically, hand-drawing is now being reintroduced in our office in order to make quick diagrams and 3-D sketch explorations (see Figure 13.7).

In 2008, Chris Walker, a computer animation artist, joined our firm. He first developed a system to search through our photos and then began to enlarge the computers' artistic capabilities. In 2009, Chris introduced the office to SketchUp, a program that turns plans into three-dimensional space and shares Vectorworks' line-edge characteristics. Suddenly, we were able to represent plant and construction materials in scale and with real accuracy. The designer could, for the first time, visually enter the virtual designed space, testing and deciding the character, material, and even the feel of the intended result. Detail renderings could still receive Photoshop refinements. These three-dimensional models could also be made directly into videos. With artistry, the light, shadow, plant growth, and even climate could be included and examined—viewed in plan and aerial views from any angle.

13.8
Collection of Digital Drawings. With artistry, the light, shadow, plant growth, and even climate can be included and examined—viewed in plan and aerial views from any angle.

13.9
Collection of Digital Drawings. With digital modelling, we were able to represent plant and construction materials in scale and with real accuracy. The designer could, for the first time, visually enter the virtual designed space, testing and deciding the character, material, and even the feel of the intended result.

In the last few years, our landscape spaces can be not only examined from within but also a high level of detail can be developed through the designer's virtual movement and observation through the space. These simulated environments can themselves represent moving water, realistic plant growth, changing seasons, shadow, and site surroundings. We can 'build' trees for specific purposes. Even the time of night or day can be simulated. Videos of movement through the design can now be seen by clients as well as designers. Simulations using these computer models, through use of Virtual Reality goggles, allow both the designers and our clients to virtually enter the design, walk around, and observe from any point of view (see Figures 13.8 and 13.9).

Video and Movement

Increasingly, clients require the leading design offices to put their competition presentations on video, which quickly inform the corporate heads about the landscape design quality of large projects. Video presentations often include "fly-throughs," "walk-throughs," and even descriptions of the design proposals by the designers.

The computer drawings lend themselves not only to short videos but also directly into videos of up to 20 minutes or more. These videos are also used to display design ideas to the public through direct-showing and also on network TV. They allow the designer to show realistic movement within and through the spaces.

Overall, I intentionally omit the high-skilled perspectives and videos done by professional production studios that are used at the end of a completely designed project and do not, at this time, directly influence our design process. I find it amazing that the refinement of representation has come so far in a relatively short period of time. It seems that the capacity grows with the graduation of each university class. So, this is the state of our art today at Peter Walker and Partners (PWP) and many other offices, and at the end of my career, but who knows what will come next.

Notes

1 Stanley White was a landscape architect and professor of Landscape Architecture at the University of Illinois from 1922 to 1959.
2 Zip-A-Tone is known as 'screentone' technique in applying patterns of texture, shade, and hatching onto drawings. Zip-A-Tone was often on a clear 'stickie-back' that was peeled and placed on a paper surface to represent texture and shading.
3 Bill Johnson was the Dean and professor at University of Michigan and co-founder of the acclaimed landscape architecture firm Johnson, Johnson, and Roy (JJR) which was founded in 1961. Johnson attended the GSD with Peter Walker and later, in professional practice, they collaborated on many projects.

14 Pieces of the World
Yves Brunier's Landscape Representations[1]

Linda Pollak

Modern landscape has been notoriously resistant to representation, dominated by a naturalistic approach whose overarching purpose is to maintain the fiction of nature's wholeness. Emblematic of this wholeness is a smooth green surface whose apparent self-evidence serves to render its meaning invisible and hence inaccessible to criticism. This smooth surface emerged in late eighteenth-century England in the landscape gardens of Lancelot "Capability" Brown, whose gently undulating lawns and lakes fashioned to look like rivers contributed to the enduring paradigm of pastoral landscape. Such timeless appearance—bearing no visible trace of the "hand of man"—was instrumental in framing a landscape's new bourgeois owners as having always "naturally" been there.

While the landscape garden was enthusiastically received in North America, its naturalistic aesthetic did not translate well to French soil. There, concurrent with Brown's practice, eighteenth-century designers and theorists married an emphasis on spatial experience with a theatrical approach to the construction of nature. Louis de Carmontelle wrote in (ca.) 1800, "Our gardens should transport us through the scenes of an Opera, we should create the illusion of a reality...."[2]

Few designers acknowledge that the representation of nature is a necessary aspect of its construction in any kind of landscape. Yves Brunier embraced this constitutive paradox in his design techniques and in his projects. His images explore and represent a nature that is both spatial experience and aesthetic object, displaced from other times and places. His images exceed by far the sum of their parts not only in their beauty, economy, and legibility but also in their emphatic lack of purity, which allows the materials of a project to coalesce within and through them to form new spaces.

A Fragment of a Greater Fragmentation

Brunier's projects disrupt landscape's taken-for-granted image-reality. By beginning with fragments of already-worked-on material, his work achieves an abstraction that is not opposed to social or physical context, and that is not a triumph of the ideal over nature, but a "passion to remake the object."[3] His approach allows representation—both in and by landscape architecture—to engage social and aesthetic spheres simultaneously, with, therefore, the capacity to address space in productive terms.

To conceive of nature in terms of fragment, and fragment as constituent of a mode of figuration, circumvents the habitual desire to break the world down into neat dichotomies of artificial and natural, form and matter, represented and real. In Brunier's gardens, all construction is reconstruction, and each element is a piece of a world in simultaneous growth and disintegration.

While the fragment as a metaphor often signifies the disintegration of a previously intact formal system, the artist Robert Smithson's conception of "a fragment of a greater fragmentation"[4] frees the fragment from the whole to allow a shift towards provisional figures, in their possible interrelationships. A person who can no longer take for granted nature's wholeness is free to perceive it as both constructed and broken and engaged in multiple local and concrete relationships. The fact that there is no whole that can be apprehended visually produces an oscillation between general and detailed readings of a landscape that contributes to unmask any naturalizing effect.

DOI: 10.4324/9781003183402-14

Michel Foucault described the garden as the "smallest fragment of the world [that] at the same time represents its totality."[5] He identified the garden as a heterotopia, that is, an "other" space, which has "the power of juxtaposing in a single real place a series of places alien to each other," with the "property of being in relation with all the others but in such a way as to suspend, neutralize, or invert the set of relationships designed, reflected or mirrored by themselves."[6]

Displacement as Figuration

Brunier's project for Museumpark[7] is made up of four distinct zones, each of which is a collage of existing and new, constructed and natural elements: the whitewashed trees of the Orchard and the anomalous, alien species of black bamboo breaking through the Asphalt Podium are elements of nature that have become figured by displacement and by a disruption of conventional syntax. Each of these elements, by being displaced from another time or space, fosters instabilities in everyday space. These instabilities interact to produce an "other" space: an oscillating field of relationships that calls into question our ways of looking at nature and culture.

The operation of figuration through displacement offers a positive key to Brunier's representations of nature in the context of everyday urban sites, where it is by definition out of place. A natural artefact within a constructed setting operates to draw the site into a relation with an "other" reality that the observer must construct on his or her own. This approach supports the representation of nature in the city, where the "presence of something, matter a community, a relief, vegetation, the sky, the earth, forces the architect into encroaching, into taking pieces off and adding new ones, never making anything from one piece of cloth and in one go…."[8]

The smaller a fragment is, the more vulnerable it becomes. Brunier acknowledged this vulnerability as one of nature's contradictory properties, writing that "the exotic is born of combinations and associations of plant families, from utilizing the evocative and the imaginative familiar but strange plants, compelling aesthetics, a kind of softness and fragility that changes perceptions."[9]

What Rem Koolhaas interpreted as Brunier's violence against nature[10] is perhaps a violence towards constructs of nature that, on the one hand, cast "her" as inherently pure and good, and underpin this purity by disallowing her to be "touched," and on the other, attempt to domesticate her in polite rows of street trees and orderly beds of planting. Brunier's nature is abundant, riotous, unstable, surreal, but also respected and given agency.

Collage as Poetic Procedure

Brunier left nothing untouched in his collages: elements of nature are torn, cut, and painted and written over, becoming, through such operations, emphatically present. The absence of deference in his willingness to layer over nature, and to layer different natures on top of each other, can be interpreted in terms of hybridization of techniques of collage and drawing with techniques of planting and constructing. Like his collages, the spaces he designed are made up of fragments of diverse evocative, sparkling, realities.

Brunier's use of collage as a poetic procedure, including his technique of drawing or painting on top of a photograph, suggests the influence of the Surrealist artist Max Ernst (1891–1971). Ernst introduced the technique of overpaint in the 1920s, as a means to register tensions between opposing values, ideas and conditions, on disparate layers of the same stratified image. He painted over images that had sunk below the horizon of cultural visibility or value—barely noticed or if noticed not valued, thus making new sense of the existing artefact.

Brunier described the site for Museumpark as a place that has sunk below the horizon of visibility: "direct, stretched out, charming, fragile, half-abandoned or partially used, and also the support for a park project."[11] He painted over photographs of the park, making new sense of the existing, through such representation.

In *The Garden of France*, an overpainting from 1962, Ernst placed at the centre a horizontal female figure, cut out from a copy of *The Birth of Venus*, a nineteenth-century salon painting (Alexandre Cabanel, 1863). In Ernst's overpainting, the figure is cradled within a multi-layered landscape, in which banded surfaces of colour celebrate a fruitful land, while two rivers flow in opposite directions around her. The painting's simultaneous two- and three-dimensional reading, the artist's preoccupation with texture, and the contradictory erotic and placid associations of elements are themes that reverberate in Brunier's artistic production. Brunier's collage-drawing of banded, internally differentiated surfaces of poplars, serviceberry trees, and concrete dunes for the *Autoroutes du Sud* project (Vienne, 1989, with Jean Nouvel) shares many qualities with *The Garden of France*.

Brunier's collages and overpaintings differ from Ernst's in their objectives, in that each image concentrates qualities of a specific proposal for a physical landscape. His transformation of the meaning of a photograph through overpainting parallels design procedures that shift the meaning of the—always already constructed—site itself. His materials are fragments, each fragment a constitutive element with the capacity to be transformed through relationships with other fragments. At Museumpark, he brought the operation of overpainting to the site, where:

> … an Arboretum of apple trees with white-washed trunks creates a welcoming feeling. These familiar fruit trees instill rhythm, their trunks scale down the existing poplars whose trunks are also white-washed.[12]

Figuring Ground

Brunier's techniques of representation—collage, overpaint, watercolour, study model—support and emphasize the construction of the ground. His representation of a ground as always already constructed allows not only for the potential disruption of a surface but also for the lack of primacy of any single surface. Nothing is taken for granted as background: each surface is a figure-in-itself, interconnected with other surface-figures, which simultaneously holds fragments together and is constituted by them. As his sketches reveal, anything can become a ground—an assembly of hundreds of chairs, a field of sunflowers, fragments of coloured materials.

The ground is where it is all happening: Brunier described a garden for the Office of European Patents as "A carpet of flowers and mosaics, … a geometrical patchwork of paths and hedges, a place of rest in a labyrinth for the eye; a surface of smells, flowers, stones, colours and levels."[13] A representation of the unbuilt memorial at Waterloo (Belgium, 1989, with Isabelle Auricoste) describes the ground as a bumpy quilt, registering the irregular topography and the patchwork of agricultural fields. For the *Autoroutes de Sud*, the ground was to be a blue intertwining of textiles, rocks, and glass. At Euralille (1989, with OMA/Rem Koolhaas), the ground of the park assumes the form of a mountain, a hub of contrasts that gathers diverse scales of urban energies around itself.

Brunier constructed the grounds of Museumpark out of disparate materials and activated these grounds through a range of techniques. In the Orchard, "a supple mineral treatment of white gravel expands the space of the street"[14] into the infinite space of the mirrored wall and vertically up the whitewashed trunks of apple and poplar trees. The social space of Zone 2—according to Brunier part vacant lot, part *teatro fantastic*—is a sturdy Asphalt Podium broken and activated by diverse nature events. Mineral meets vegetal in the ground of Zone 3, the Romantic Garden, where confetti-like constellations of scattered brightly coloured bricks echo the confetti of fallen leaves, alongside a field dense with flowers, with an artificial river of stones and sparkling glass balls below. The ground of the fourth zone is a terrace of hard and soft surfaces that enters the Kunsthal (OMA, 1992) and winds its way up through the building, as a broad ramp, effectively appropriating the museum as part of the park.

Representation of Plants

The overlapping layers of space in Brunier's landscapes often correspond to the growth habits and spatial properties of plant materials. He juxtaposed plants with each other and with mineral elements in combinations of position, scale, and colour. His pencil and watercolour sketches detail encounters between heterogeneous elements such as the formation of a space by a fringe of intensely coloured overhanging leaves above and a reflective ground plane below.

As with his attention to the ground, Brunier's engagement with and figuration of plant materials attains a high degree of specificity. Evocative descriptions of plants accompany the drawings and plant lists for Museumpark, moving from the "delicate" to the "vaporous," from "confetti patches" to a reddening island, and a "fan of greenery and sunlight."[15] The planting plan of the Romantic Garden in Museumpark is an exuberant painting of red, yellow, green, blue, and orange comprising over 50 species of plants. As accompanying collages indicate, each season lays down another set of colours.

Brunier's verbal descriptions of plants are as evocative as his images. At the restaurant of the Hotel St. James (with Jean Nouvel, 1987) where "the chef puts dishes together like colourful paintings," Brunier designed a sensual garden that allows "existing romantic plants [to become] electrified by their relationships with new colours… and a vineyard, like an architecture of plants … [that] rubs against fluorescent rows of white brambles, osier, and red lime trees."[16]

Writing about the Park at Euralille, Brunier stated that:

> … a different world has been created, based on the differentiation of constant surprising plant matter. Emotions surge with the rhythm of the seasons, the colours and scents of flowers in bloom, the abundance of fruit and the unusualness of bark, weaving a story in which all may find pleasure because it tells of the generosity of nature.[17]

Legacy

Yves Brunier's artistic practice can be likened to that of Robert Smithson in its unsentimental collaboration with "chance and change in the material order of nature"[18] and in its lasting impact on how we understand the representation of constructed landscapes, notwithstanding the tragic curtailment of both of these men's lives.[19] In discussing the possibility for a radical experience of landscape space, Smithson wrote that "[w]hat we take to be the most concrete … often turns into a concatenation of the unexpected. Any order can be reordered."[20] Brunier's construction of "different worlds" disrupts established meanings, bringing past and present together in new formations in relation to physical and social structures in the world.

My copy of Brunier's monograph has allowed me to share his work with generations of students. His collages engage a specific register that is relevant to a pedagogical agenda of shifting emphasis away from an object-oriented focus towards a more inclusive attention for environments. The specifics of his collage techniques, including the way that they establish fields in tension, have a particular value insofar as they redirect certain processes of collage that architecture, with its dependence on orthographic representation, has often flattened.

In my *Landscape Theory* seminar at The Cooper Union, my objective is to inspire and support architecture students to integrate landscape thinking and design in their own creative practice. In studio, I encourage designers to sketch by hand, including with coloured pencils, and to make composite representations that include physical as well as digital drawing, photographs, and different kinds of writing. In the seminar, Brunier's watercolour and pencil sketches and collages inspire architecture students to make their own. As Octavia Parker, who studied Brunier's images by overpainting them in 2017, writes:

> Yves' work is beyond ornament in/as landscape. His landscapes serve as live activity systems. Stills. Analyzing his work gave me the confidence to explore different representational techniques in my own work, specifically within building analysis. His Museumpark project was my introduction into landscape. I think through collages of forms and colours with respect to movement when I construct landscapes.[21]

Figures 14.1–14.6 are some of Brunier's representations of Museumpark.[22]

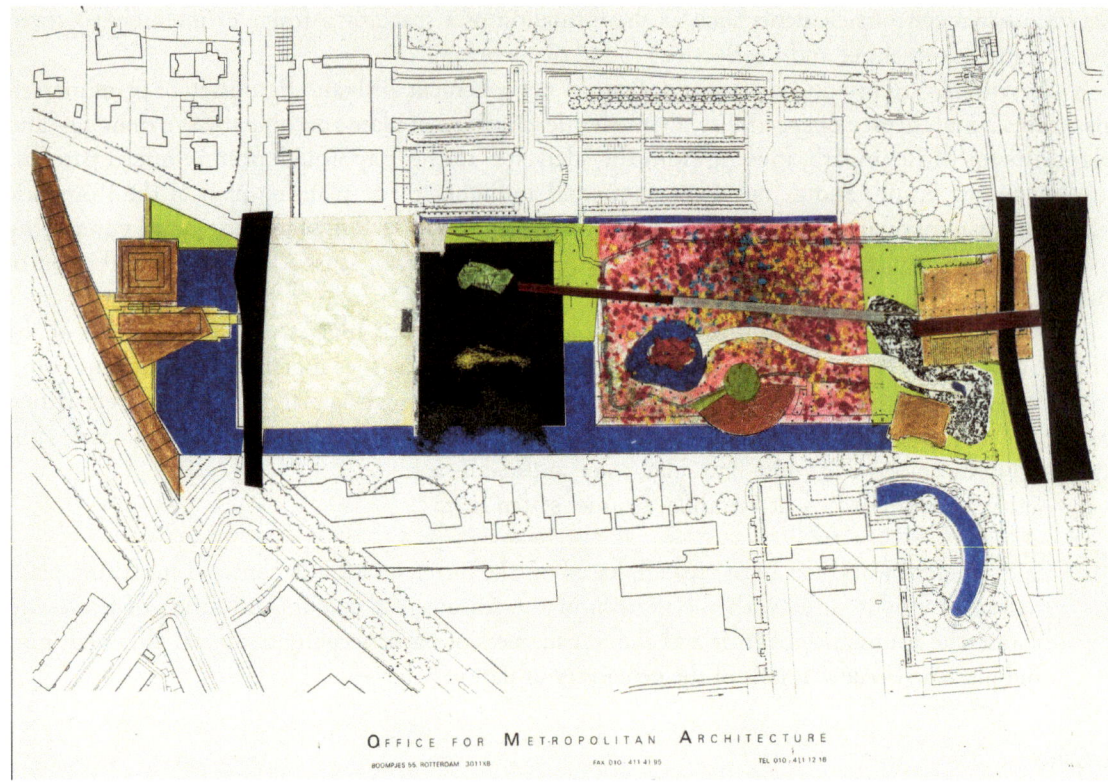

14.1
Collage plan of Museumpark.

14.2
Collage/overpainting of Museumpark Orchard.

Pieces of the World

14.3
Overpainting of Museumpark Orchard.

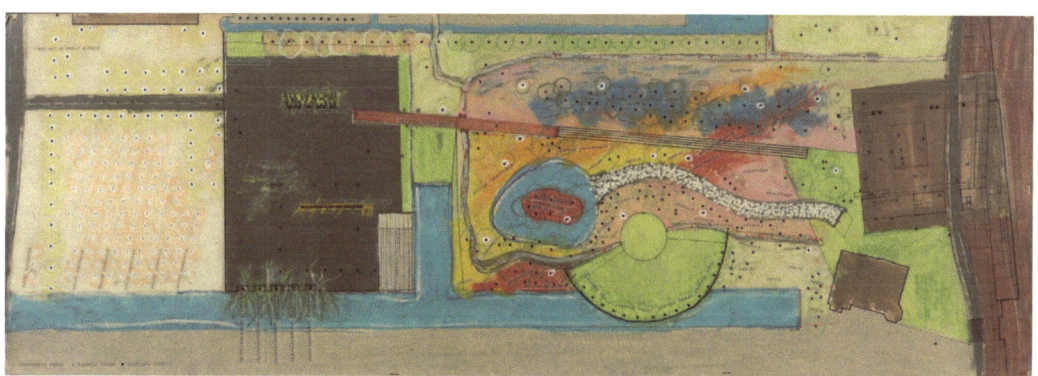

14.4
Plan study of Museumpark materials and planting.

14.5
Collage/overpainting of Museumpark Romantic Garden in summer, with bridge through the flowers at left.

14.6
Overpainting at Romantic Garden in summer, with white-flowered vines extending up the trunks of existing trees.

Notes

1. This essay opens up and expands upon a text I wrote in 1999 for the Storefront for Art and Architecture exhibition, "Yves Brunier: Landscape Architect," co-curated with Paula Meijerink.
2. Weibenson, D. *The Picturesque Garden in France*. Princeton, NJ: Princeton University Press, 1978, p. 97.
3. Krauss, R. *The Optical Unconscious*. Cambridge, MA: MIT Press, 1995, p. 11.
4. Smithson, R. "A Sedimentation of the Mind: Earth Projects." *Artforum*, 1968, 7(1): 50.
5. Foucault, M. "Of Other Spaces." trans. J. Miskowiec. *Diacritics*, 1986, 16(1): 26.
6. Ibid., pp. 25–26.
7. Following his death in 1991, Brunier's design for Museumpark was carried by Petra Blaisse's firm Inside/Outside, with OMA. The park was restored by Inside/Outside with OMA (2008–2011), including changes to some elements.
8. Agacinski, S. "Sewing Machine: Building Monumentally." In: P. Osborne and A. Benjamin (Eds.). *Thinking Art: Beyond Traditional Aesthetics*. London, UK: ICA, 1991, p. 215.
9. Brunier writing about the Park at Euralille, in: Claramunt, M. *Yves Brunier: Landscape Architect Paysagiste*. ed. M. Jacques. Bordeaux, France: Arc en rêve centre d'architecture and Birkhauser Verlag, 1996, p. 106.
10. Ibid., p. 89. (Interview with Rem Koolhaas, 1996).
11. Ibid., p. 106.
12. Ibid., p. 106.
13. Ibid., p. 116.
14. Ibid., p. 106.
15. Ibid., p. 106.
16. Ibid., p. 114.
17. Ibid., p. 114.
18. Smithson, R. "Frederick Law Olmstead and the Dialectical Landscape." *Artforum*, 1973, 11(6): 63.
19. Smithson died in a plane crash in 1973 at age 36. Brunier died of AIDS in 1991 at age 28.
20. Smithson, R. "A Cinematic Atopia." *Artforum*, 1971, 10(1): 53.
21. Octavia Parker, email to author, April 12, 2021.
22. These images are provided by Het Nieuwe Instituut in The Netherlands.

References

Agacinski, S. "Sewing Machine: Building Monumentally." In: P. Osborne and A. Benjamin (Eds.). *Thinking Art: Beyond Traditional Aesthetics*. London, UK: ICA, 1991, pp. 209–215
Claramunt, M. *Yves Brunier: Landscape architect paysagiste*. Ed. M. Jacques. Bordeaux, France: Arc en rêve centre d'architecture and Birkhauser Verlag, 1996.
Foucault, M. "Of Other Spaces." Trans. J. Miskowiec. *Diacritics*, 16(1) (1986): 22–27.
Holt, N. (ed.). *The Writings of Robert Smithson*. New York, NY: New York University Press, 1979.
Krauss, R. *The Optical Unconscious*. Cambridge, MA: MIT Press, 1995.
Smithson, R. "A Cinematic Atopia." *Artforum*, 10(1) (1971): 53–55.
Smithson, R. "A Sedimentation of the Mind: Earth Projects." *Artforum*, 7(1) (1968): 44–50.
Smithson, R. "Frederick Law Olmstead and the Dialectical Landscape." *Artforum*, 11(6) (1973): 62–68.
Weibenson, D. *The Picturesque Garden in France*. Princeton, NJ: Princeton University Press, 1978.

15 Hands On!

Petra Blaisse

Landscapes are primal scenes of promise and mystery. They cater to an extension of the moment by sprinkling the present with tantalizing hints of sensory immersion—smells of a memory in the making, a rush of wind that tickles the world to move in slow motion, or soundscapes that add depth to the environment. Multiplied by the many viewpoints and perspectives at hand, navigating the landscape's different settings, paths, topographies, and edges creates a sequence of moments to be collected like meaningful treasures. The experience of a landscape over time is one where changeability overrules predictability. Thus, the same site offers a succession of disparate instances in which the flow of time is made tangible. Anchoring into the present moment is also a way for the natural world to remind us of an unfamiliar future. Landscapes cater to notions of age-old geological processes, alternative life spans, and timelines that far surpass our presence within the world.

Despite dealing in ever-changing settings or inconceivable temporalities, landscape architects nevertheless take up the precarious position as predictors. That said, there is a fundamental misfit between the aura of landscape and the tools that we are currently using to evoke it. Noel van Dooren and Anders Nielsen write that representation in landscape architecture is often preferred in absolutist terms and is expected to provide a "stand-in" for a future reality.[1] In recent years, representation is venturing towards increasingly hyper-real renderings that are rushing towards a palatable endpoint that, in reality, does not exist. To reach a "true to life" representation, it is important for it to be aligned with the irregular rhythms and affective experiences of a landscape, making *approximation* a better fit than a *prognosis*. This chapter explores how such an approximation is achieved through abstraction, through representations that play into the imagination and incite scenarios appropriate to the living, breathing settings they communicate. In doing so, this text acknowledges that with great representation comes great responsibility. As much as the profession might benefit from appealing to the imagination, designers carry an obligation to their audience, clients, or the context within which they work to communicate with efficacy and resemblance.[2] This chapter provides a series of commentaries that address different techniques across a spectrum of representation—from hyper-realistic to abstracted forms—to explore the extent to which these fulfill the demands of the context of their creation.

Landscape design projects commonly follow the prescribed architectural phases in their development from sketch and schematic design to design development, tender, and construction phases—a traditional logic. This rationale is up for scrutiny, as landscapes, unlike architecture, do not offer a precise and predictable outcome. Landscape design is about creating the conditions within which the landscape can eventually flourish. But, in practice, the landscape is expected to be presented in the same language as architecture, as if it is mainly a built world with some plants and trees on it. The language is not to dictate the intention, but rather the other way around: the vision comes first, the technique follows. Furthermore, landscape design development does not always follow the same phasing as in architecture. The landscape might be realized ahead of a larger situation that is yet to be developed. It might require more time to negotiate the right conditions within the architectural or urban construct, or it could act as the final addition to an existing context as a kind of sustainability infill. Neither the endpoint nor the process by which landscapes come into being fits into a language that often favours a prescribed format.

Landscape representation is always an illusion, a vision, an intention made visible. It is never about reality. It is not a static structure, a building. It is alive and inevitably susceptible to changing circumstances such as fluctuations in climate, seasonal change, extreme soil conditions, and developments in the direct surrounding that can only be partially—if not minimally—controlled, let alone predicted. For representation to precede actual existence, it can therefore only be an informed guess. No matter the precision offered through 3D renderings, digital collages, and AutoCAD drawings, the real outcome will be different. Reality is only revealed on the site in question, both when the time comes and with time itself. Faced with an astonishing divergence of possible outcomes, the current bias that the representation of landscapes functions as a calculable domain is in itself unrealistic. Abstraction, instead, serves as a better reflection of the initial intention and the future reality as opposed to photo-realism.

Human-made landscapes address and create a sequence of green spaces intersected by meandering or linear trajectories that lead the viewer through or along. They offer various sensations, impressions, and experiences that change with the seasons and over the years, depending on the viewer in question. While landscapes do sometimes include built structures such as retaining walls, paths, water basins or swimming pools, drainage systems, pavilions, grottos, or follies, which can be drawn out meticulously, they primarily consist of a living green of different heights, structures, colours, shapes, and levels of transparency; flowering, fruit-, nut-, or seed-bearing, evergreen or deciduous. These listed components can certainly be shown in a representation, but only at one specific moment in time. The light-green, sprinkling forest with its carpet of blue blooming bulbs in spring, the rich flowering meadow in summer, the plant border in winter, and the view over the pond of a misty hill on an early morning—all of these are fugitive entities that exist only in the interim.

Ever since the development of digital drawing and representation tools, almost all parties—among which are our colleagues—now expect sketches, spontaneous ideas and epiphanies that take form from our brains to our hands, to be translated into digital collages and 3D renderings. While an AutoCAD plan drawing can bring useful insights into placement, measurements, and logistics, it does not teach us anything about the experience of the garden, nor does it offer an understanding of its scale or spatial composition. The transformation into the digital has had an alienating influence on the choreographer's actual connection to the space they have envisioned. For instance, often, if a colleague is asked to pinpoint the effect of a dimension on an area or particular object or to recall a dimension of a thought-out project by heart, there would be no answer that comes to mind—it would have to be looked up on the screen. This is a disenchanting after-effect to the immediate impulses that guide a vision in its earlier stages.

Yves Brunier would photograph the existing site and make black-and-white copies from these photos. He would then use these copies as a foundation to start drawing, painting, and sometimes even gluing little scraps of coloured paper onto them to mimic an explosion of flowers. Through this simple contrast, he added verve to his vision. He constructed rough models with papier-mâché for topography, pieces of sponge for shrubbery, and earplugs for trees. The experimental use of material brought excitement in itself, an enthusiasm that could be channelled into dreaming up a livelier and more beautiful place.

Piet Oudolf draws numerous multi-coloured blobs, stars, and dots on paper with well-sharpened crayons on chalk-paper, with each of the forms representing a specific perennial, grass, bulb, or shrub. He does not sketch nor draw a 3D representation of a future garden but shows references to his earlier gardens and draws planting plans within a composition of plant areas that define paths and routings in relation to the buildings or adjacent landscapes.

David Hockney paints fantastic landscapes in subjective colours. His representation is detached from reality but invites viewers into an energetic world full of life and movement. I have made 'hand-made' collages out of newspapers, magazine cuttings, found pieces of paper, or cardboard scraps to represent moods, compositions, atmospheres, types of usage, colour, form and spatial arrangements, and materializations. I sketch in search of the right direction for a design to develop towards, to clarify an idea, to compose a diagram, define the correct representation of trees or plantings, to symbolize differences in structure, colour, or seasonal effects. None of these representations depict or have an obligation to reality. We build models of paper, cardboard, nails, plastic, textile, beads, sweets, needles, and wooden scraps and left-overs.

Figure 15.1 was one of the six collages composed for a new state prison landscape design in the Netherlands during the late 1990s. The conceptual framing of the garden is an assemblage of images gathered from different sources. The layering of separate aesthetics into a surrealist, atmospheric whole evokes multiple settings in a single representation. As the language of this representation surrenders certitude to the realm of imagination, it broadens its scope and starts to contain the diversity that a landscape possesses.

A similar dynamic happens in a collage created for one of the two garden designs of the Stedelijk Museum Extension competition during the early 1990s (Figure 15.2a and 15.2b). The collages depict the rich layers that one can experience from within a garden that, once represented from above, appears a seemingly simple, graphic composition. Due to its layers, the collages, in all their texture and depth, look like the sum of a series of snapshots that could have been extracted from different times of the day or separate seasons altogether. It synchronizes a sunset with a foggy morning, as seen in the mirroring wall that lines one flank of the enclosed garden and even addresses the rising water level specific to the site. It captures all kinds of changes across incommensurable timelines. In so doing, a story is created across space and time, as several moments in time are discerned simultaneously.

Figure 15.3 is a quick sketch of the landscape design for the Hammer Museum, Los Angeles, 2000. This image visualizes the 'patchwork' landscape with the placement of trees. The trees are quick, stick-like gestures to suggest where they could be placed. Figure 15.4 represents a typical experimental model that we create in our office. This model was created in collaboration with Ai Wei Wei (and his cats) in Beijing. It represents the first series of studies done for the Walker Art Center landscape in Minneapolis, designed by our firm, that was to include an integrated intervention by the artist.

In Figure 15.5, two garden designs were presented parallel to the architect's two options for the floor plan of the extension of the Stedelijk Museum's main exhibition space. The study model shows an "art garden" that could be taken in and considered as an art piece in itself when seen from the spacious terrace that hovers above the garden, and an "edible garden" that is best encountered from within, where every plant and flower could be consumed.

Nevertheless, my team and I are forced to utilize digital tools in our practice. We try to include the 'experimental' and 'childlike' quality of the previously mentioned techniques by, for instance, imitating the naiveté of the hand drawing or even adopting children's book style (Figure 15.6).

15.1
Collage for one of the six gardens for a new state prison in Nieuwegein, the Netherlands, 1997.

15.2
(a) and (b) Collage for one of the two garden designs for the Stedelijk Museum Extension competition, 1992.

15.3
Sketching the landscape design for the Hammer Museum, Los Angeles, 2000.

15.4
Experimental model made in collaboration with Ai Wei Wei (and his cats) in his studio in Beijing. It represents the first series studies completed for the Walker Art Center landscape in Minneapolis, designed by Inside Outside that would include an integrated intervention by the artist.

15.5
Study model for the "Edible Garden" version of the Stedelijk Museum Extension competition, 1992.

15.6
Rendering showing the activated, landscaped wall to overcome all height differences and invite people, for the competition design for the Roshen Chocolate Factory in Kiev, 2016.

Other times, we add a 'girly,' almost mocking touch to our renderings or, to my dismay, collaborate with professionals in computer graphics, only to be left with the feeling that we sold our souls. Or, we exaggerate an educational purpose by creating digital collages that loudly communicate themes of sustainability, which is especially urgent in the urban realm (Figure 15.7).

To conclude, an aspect that I do love about digital representations is the technical construction drawings (Figure 15.8). The more layered, complex and meticulous, the better, as it shows the extent to which the creative process can turn into a physical reality. As a designer, your involvement is

15.7
Rendering with an educational purpose. As part of Inside Outside's landscape advisory role, this is a drawing of a municipal landscape competition inside a new, dense building complex in Utrecht, 2021.

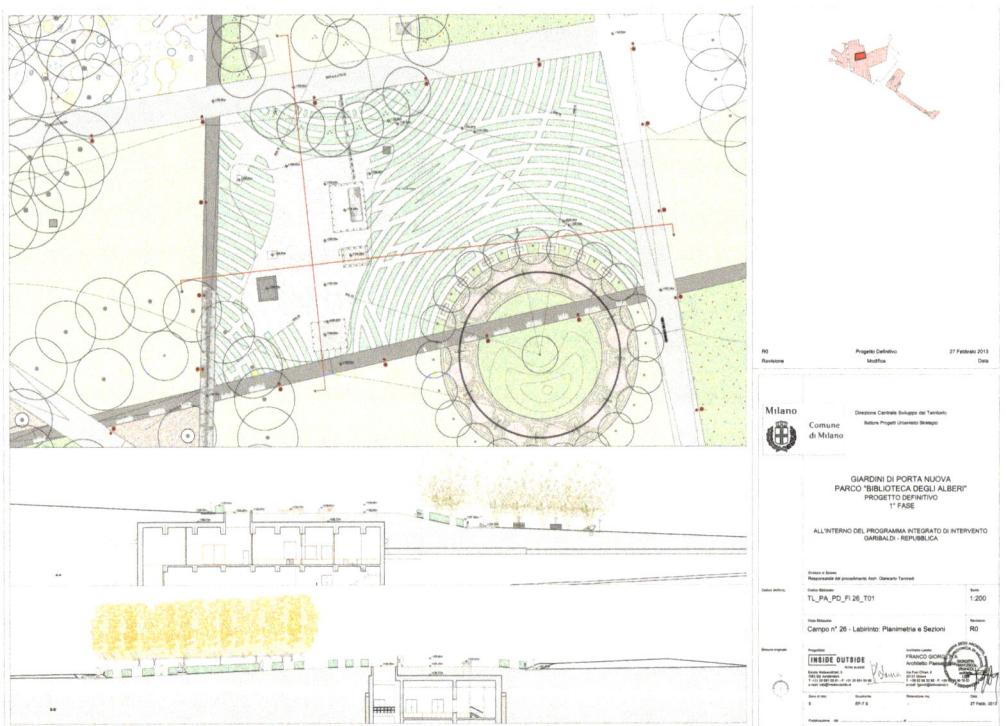

15.8
Plan and section depicting the underground space upon which the park was to be realized. For the Milanese park, Biblioteca degli Alberi, for which the competition design was commissioned from 2002 to 2003 and eventually under construction between 2010 (the start of the commission) and completed. The opening date was October 2018.

ultimately guaranteed until the final, technical detail that can safeguard a healthy condition, a base. From there on out, the reigns on this sophisticated system need to be released.

Notes

1 Van Dooren, N. and Busse Nielsen, A. "The Representation of Time: Addressing a Theoretical Flaw in Landscape Architecture." *Landscape Research*, 2008, 44(8): 997–1013, p. 997.
2 Kingery-Page, K. and Hahn, H. "The Aesthetics of Digital Representation: Realism, Abstraction and Kitsch." *Journal of Landscape Architecture*, 2012, 7(2): 68–75, p. 68.

References

Van Dooren, N. and Busse Nielsen, A. "The Representation of Time: Addressing a Theoretical Flaw in Landscape Architecture." *Landscape Research*, 44(8) (2008): 997–1013.
Kingery-Page, K. and Hahn, H. "The Aesthetics of Digital Representation: Realism, Abstraction and Kitsch." *Journal of Landscape Architecture*, 7(2) (2012): 68–75.
Zeng, Y. "Observation and Representation: On Recognition and Expression of Natural Sites in Landscape Architecture." *Landscape Architecture Frontiers*, 7(5) (2019): 10–23.

16 Freedom from an Innocent Landscape

The Visual Communication of West 8

Adriaan Geuze

Rooted in the Netherlands, design firm West 8 began as a shared commitment to the heritage of the country's landscape: the tradition of land-making against the natural forces. This engineering logic is intertwined with the practice of documenting, illustrating, and envisioning landscapes both through pictographic and analogue techniques as well as the embedded context of location and use. This complex, multi-disciplinary approach to land-making is in the firm's DNA. Between art and engineering comes the manipulation of landscape—it's layered, honest, and systematic.

West 8

Situated in the docklands of the port of Rotterdam, at the confluence of the rivers Maas and Rhine, West 8 was founded in 1987, and has grown from a small team of four squatting within the city's iconic Hotel New York, to now occupying three offices across the Netherlands, Belgium, and North America. With a multi-disciplinary team of 70 designers from 18 different countries, the team works as a collective to remain at the international forefront of design, with a diverse portfolio of executed projects all over the world.

The Legacy of Constructed Landscapes

With the turn of the seventeenth century, the Dutch began to solidify their presence in landscape representation. The nature of the Dutch landscape is artificial and structured, and above all: flat!

Inherited from the Italian renaissance, the Old Masters shifted the viewpoint to represent the horizon line from the perspective of the human eye and in doing so, elevated the beauty of the engineered landscape in a new way. The landscape was manipulated and dredged out of water for military defences, waterways, farmland, and international trade.

The necessity for documenting these 'new territories' became of equal value to representing the existing landscapes, and when artists were first asked to represent them, they discovered the power of the painter's hand—the power of editing. The placement of an iconic tree, windmill, tower, or ship at the straight horizon often acted as a touchstone for the collective mind to recognize the landscape immediately, but at the same time, there was little hesitation to adjust, abstract, or enhance reality for the benefit of composition. Decorated with showers of light, magnificent clouds, or exaggerated foreground/background layers, these works were produced at an immense scale, creating hyper-realistic paintings the viewer could imagine walking directly into.

These techniques frame the backbone of West 8's approach to the landscape: human-made landscapes, both real and abstracted, layered with cultural iconography to evoke a response in the viewer.

The Innocent Landscape

The urban park was a nineteenth-century concept, its invention necessary to provide relief to the urban victims of the new, untamed metropolis. Unlike the grand boulevards with which Haussmann carved monumental green arteries into the existing urban fabric, and the early urban parks that designers fitted into the grid, the later parks actually determine the unbuilt city. The large Paris parks of the late nineteenth century, Bois de Boulogne and Bois Vincennes, and Olmsted's big North American parks were the original models of the principle through which landscape guides urban development. Greenery delineated the new edges of the neighbourhoods to be built, and the green investment was immediately repaid by increased land value. Planning, real estate development, and the poetic presence of nature were combined. Properly regarded, these were the purest forms of landscape urbanism or landscape-as-infrastructure. Landscape is inherently poised as a medium to undertake many of the world's problems: variable and powerful enough to lead the urban realm with true place-making. However, the expectations of filtered, cosmetic images of a sanitized landscape—often the requirements of community participators, developers, government representatives, and other stakeholders—can create a demand for visions which lack the contrasts, challenges, and realities of everyday life. Is it inevitable that Landscape Architecture should become restricted to be only about cultivating 'innocence'?

This Olmstedian principle seems still to be the ideal of landscape urbanism, although, in practice, hardly any critical attention is paid to some of its weaker aspects. Why is it so easily accepted that the green of parks will always bring a better world?

First, the steadily increasing area of overdesigned, suburban green structures is of a dubiously hybrid character: searching for the cultural significance of beloved nineteenth-century city parks; but on the other hand, they attempt to create an idealistic wilderness. Realization often results in a strange non-world of cultivated innocence. The essential characteristics a park needs to survive, so exhaustively described by Jane Jacobs, are almost always lacking. According to her analysis, for parks and greenery to succeed, a good context is fundamental. Many city dwellers see peripheral green zones not only as valuable green background but also as potentially dangerous, and as places to be avoided. There is simply too little activity and no mixing of user groups. Park designers have not succeeded in giving these parks the allure of nature and wilderness.

Second, landscape architecture is fundamentally linked to nature, to mother earth. But the perception of "nature" has largely remained explicitly cultural, quite different from one another, each with its particular perceptions of nature. When you talk with different nationalities about nature, you are confronted by deeply rooted feelings and cultural convictions, all of which are assumed to be a matter of "common sense."

Finally, the pretension often is that parks are the result of ideology and craftsmanship and are therefore inherently unique and valuable. However, landscape architecture, in contrast to architecture, is concerned almost exclusively with the public realm—parks, boulevards, riverfronts, streetscapes, and so on. To reach decisions and establish finances, the designer must work with politicians, local citizens, and bureaucracies with diverse legal systems. Landscape architecture will always focus on outreach, public opinion, interaction, public policy, implementation, and compromise. The discipline cannot avoid responding to socio-political contexts. Economists have an acronym to identify the forces driving development: PESTEL (politics, economics, sociology, technology, environment, and law). It is crucial that contemporary planning initiatives explicitly take these factors into account. Clearly such diverse issues as governance and legislation, high- and low-tech implementation strategies, grassroots advocacy, and megaprojects all are attendant on public policy.

In practice, landscape architects and park designers work in a realm between illusion and public policy, and the work is inevitably the most banal and compromised among the design disciplines. At the end of the day, are the built realities anywhere close to the dreamt-of parks and artist's impressions?

The Alignment of West 8

In the late 1980s, before the use of computers in the design process, 'diapositive' photography and drawings on acetate were the most effective means of (literally) projecting a visualized landscape.

At this time, photocopy machines were the peak of innovation, and there were seemingly two paths of landscape design: obeying the Beaux-Arts or exploring new technologies.

As a young office, West 8 brought together a synthesis of both tribes, utilizing the methodology of collage, composition, and repetition with the inherent distortion from the process of cutting, pasting, and copying. West 8 inserted itself into the friction: composing and playing with context, horizon line, and the commonly accepted architectural language of 'black and white' (see Figures 16.1a and 16.1b, 16.2a and 16.2b, and 16.3a and 16.3b).

The image-making process of the time involved searching through existing media of magazines and newspapers, finding the right silhouette, figure, or texture to fit the scene. Enlargement was only possible via a scale factor, with every reprint aiding the contrasts between black and white ink. Accessibility proved another limitation, as the entire image-making process had to be completed while traveling back-and-forth to the only Xerox Photocopy Machine available for public use, at the World Trade Center in Amsterdam.

Playing within the studio environment, West 8 combined the use of physical maquettes with creative photographic techniques; wide-angle lenses created exaggeration, illuminations were used to enhance, photographic negatives were dissected and overlaid, and elements were omitted. These methodologies altered, enhanced, and transformed the image for the viewer. The outcomes of these experimental representation techniques still break through boundaries today. Is it art? Poetry? Playful or academic? In the end, the 'ensemble' became an art form in itself.

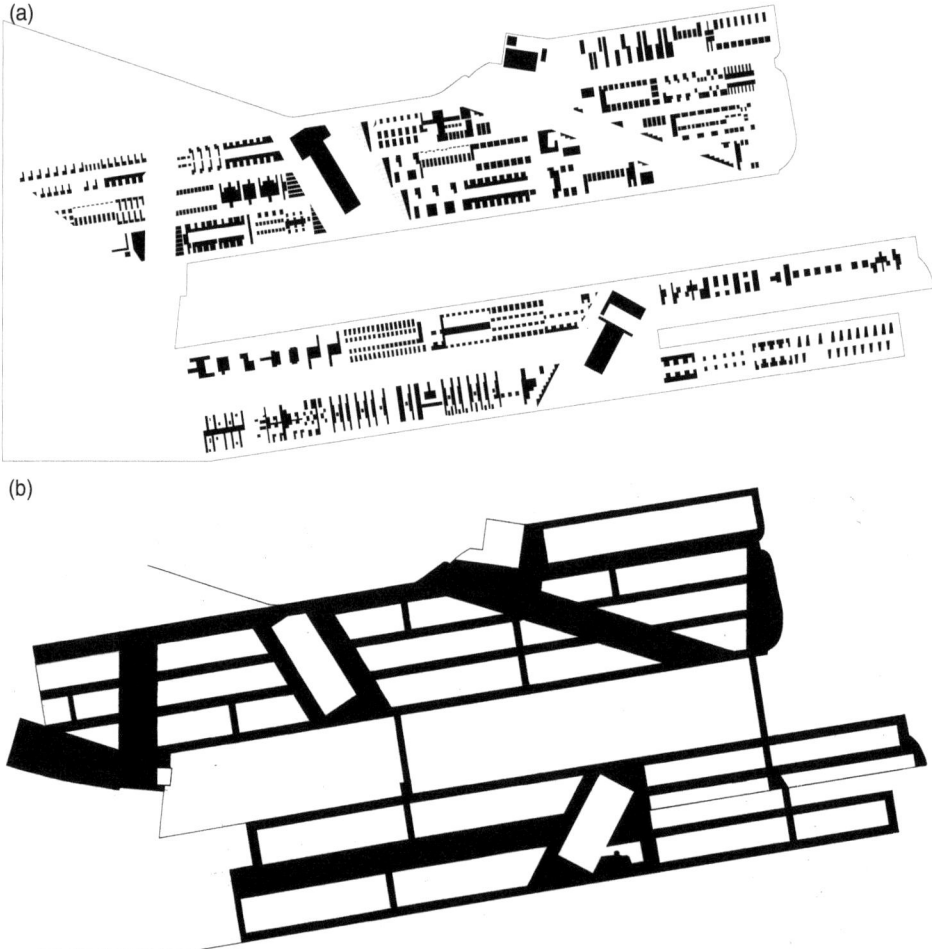

16.1
(a) and (b) Borneo-Sporenburg, Amsterdam, the Netherlands Figure/Ground. Solid and Void Representation, taking inspiration from the 1748 Nolli map, La Nuova Topografia di Roma.

Freedom from an Innocent Landscape

16.2
(a) and (b) Borneo-Sporenburg, Amsterdam, the Netherlands Xerox Perspectives. The power of distortion and scale shown through cut-and-paste collages. First found textures, then assembled, and Xeroxed as black and white copies.

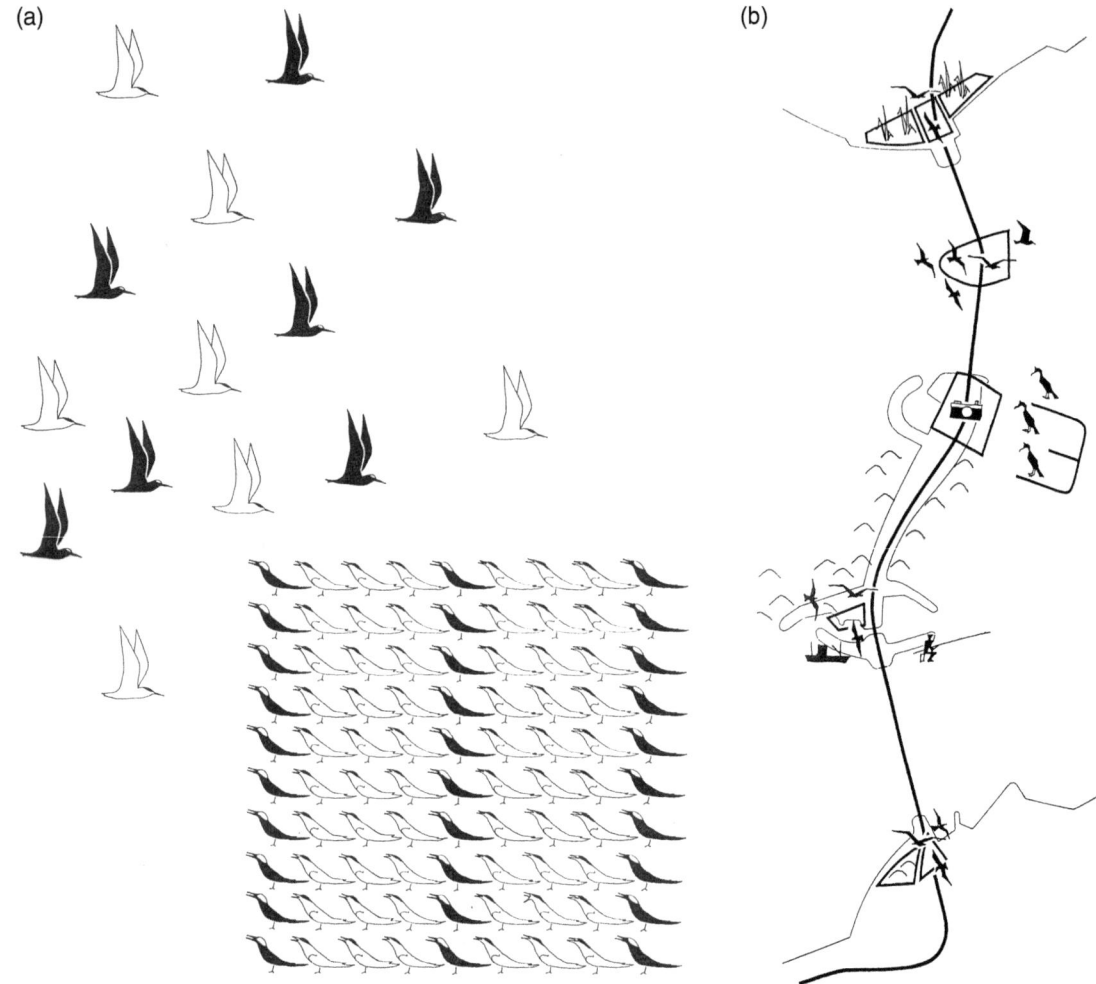

16.3
(a) and (b) Eastern Sea Storm Surge Barrier (Oosterschelde), Zeeland, the Netherlands. The photocopier as an image-making tool. Black and white birds cut, paste, and copied via a Xerox Copier. Alongside: Diaromatic site plan, annotated with contextual layer of the intervention's intended usage by people and birds.

The Layered Approach

The days when ideas of landscape were displayed with a plan and section, only digestible by those select few educated in the field to compose an environment in their minds, are long-gone. The ability to create documentation in a quicker, more accurate fashion has allowed for more 3D visualizations, wireframe exports, and DXF formats as well as the expectation to deliver a whole taxonomy of options and schemes.

The rapid evolution of computer-aided design escalated innovation and provided designers with radical accuracy and total clarity of their ideas. However, as the Dutch masters knew, there is power in distorting the existing image to create an intrinsic relation far greater in value than a picture-perfect representation. Freeing the viewer from this narrative enables the complexity to truly shine through. Perfect, clear, sanitized landscapes are not real—they do not create joy. Without any story to uncover, the viewer is likely to remain un-captivated and unsatisfied.

West 8 was born from the Dutch landscape tradition, where all nature is a concept, and where all nature is constructed. This is not merely landscape as a product of engineering design, a leftover-space of architecture, of straight lines and concepts—but a landscape that is the outcome of a systemic and multi-layered approach. With a legacy of farmers, surveyors, monks, and engineers, West 8 is interested in the narrative semantics that a visual composition may contain. In search of poetry, using abstraction and illusion to create a more accurate reality—a hyper-reality (see Figure 16.4a and 16.4b).

Freedom from an Innocent Landscape

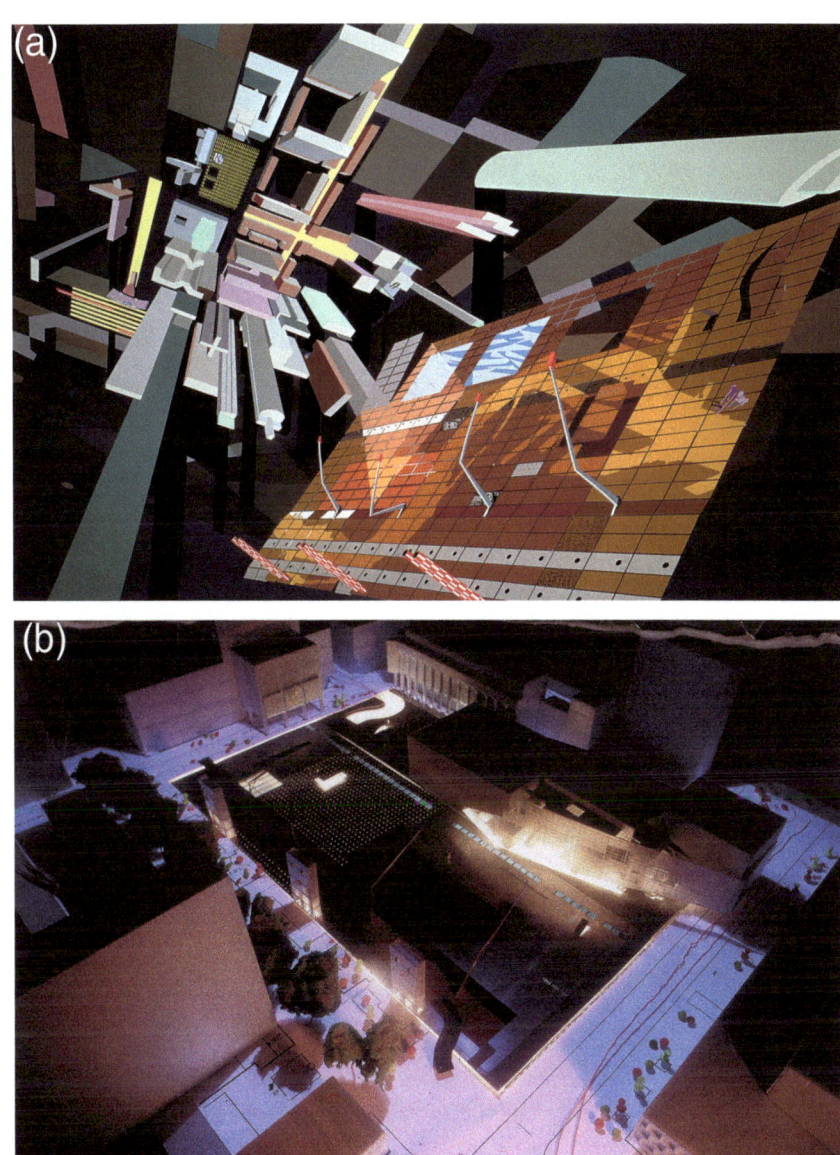

16.4
(a) and (b) Schouwburgplein, Rotterdam, the Netherlands. The combination of physical maquettes, lighting and photography techniques copied, overlaid, and recreated with individual pieces of coloured paper. Alongside: Photo of illuminated maquette, used as a base.

Strategic Editing: Walking the Line between Honesty and Deliverability

What can be perceived as a total abstraction of reality is, instead, the output of a systematic response to contextual clues. The image becomes an artefact to be decoded by the viewer—a place they can picture themselves within, a larger-than-life painting for them to walk into. The power of utilizing cultural touchstones as a cognitive factor relates to the concept of identity: where people's minds recognize visuals that represent a specific culture or identity belonging to the place they inhabit. People are rooted in contexts, landscape is nested in place, and genius-loci is geo-referenced. These techniques of representation, abstraction, distortion, and layering can be seen across all of West 8's output.

Representation

To evoke the heritage nature of landscape design for the common good of the people, West 8 utilized the language of an Olmstedian-style site plan for the masterplan of Governors Island (Figure 16.5a

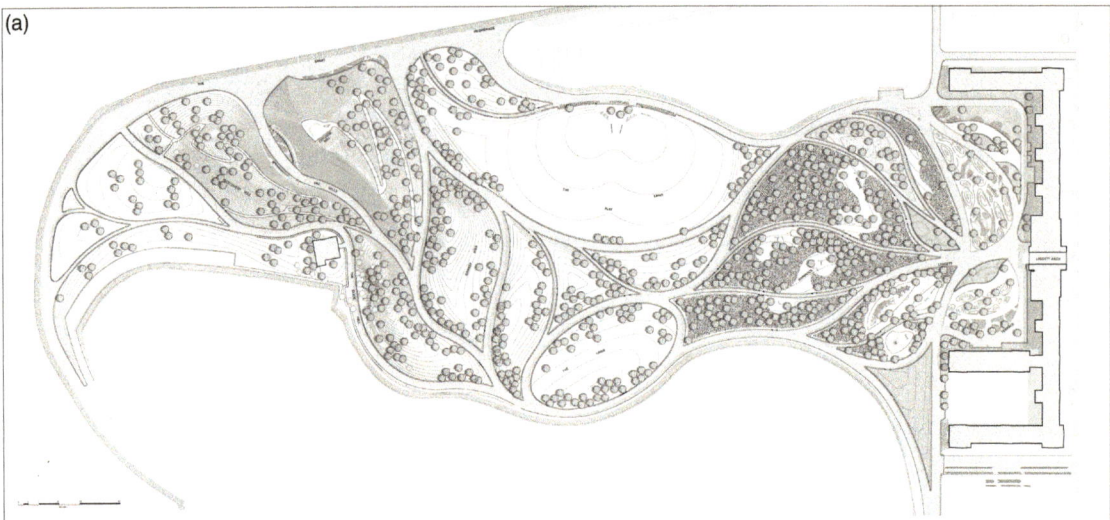

16.5
(a) and (b) Governors Island, New York, USA. Plan, black and white artistic CAD plan, reflecting language of an Olmstedian-style site plan for the masterplan of Governors Island. Carefully rendered perspective overlooking the Hudson River, wide panorama view.

and 16.5b). A duality within one drawing—both recognized as an inviting, welcoming space that lends itself to be explored by the visitor and a continuation of the many layers of heritage within the site. West 8 were also very interested in the dramatic context of the Grand Egyptian Museum in Cairo, positioned at the edge of the Giza Plateau with the generated geometry from the sightlines towards the pyramids. This new global icon required a multi-layered understanding, a systematic approach across micro and macro-scales to connect the masterplan-level infrastructure logic and the eye-level of the visitor experience. In this 'XXL project,' no doubt, the West 8 team decided to choose an almost engineer-like representation of the plan. A super heavy file, working through the scales, being blunt and explicit (see Figure 16.6). In West 8's image from the Toronto Waterfront masterplan (Figure 16.7), you will recognize two ice skating players in the frozen winter on Lake Ontario. The objective was to create an image that you can tell, understand, or identify immediately, without any title.

Distorting

The identity of a location is inherent in the way West 8 represents a project. It is not an imposed language, but rather developed as an outcome of the context of the intended audience. Instead of delivering hyper-realistic, glossy renderings for the Yongsan Park project in Korea, images for the design were rooted in the Korean tradition of painting. Combining the accuracy of computer-modelling with the ancient artistic techniques of 'aquarelle,' pencil colours, papyrus, and silhouettes of light let the viewer imagine and connect to these images as a place where they want to be (see Figures 16.8a and 16.8b and 16.9a and 16.9b).

Comparatively, on Governors Island, the more antagonistic representation is a provocation to generate collective awareness in the mind of the typical New Yorker. The project's iconic poster shows

Freedom from an Innocent Landscape

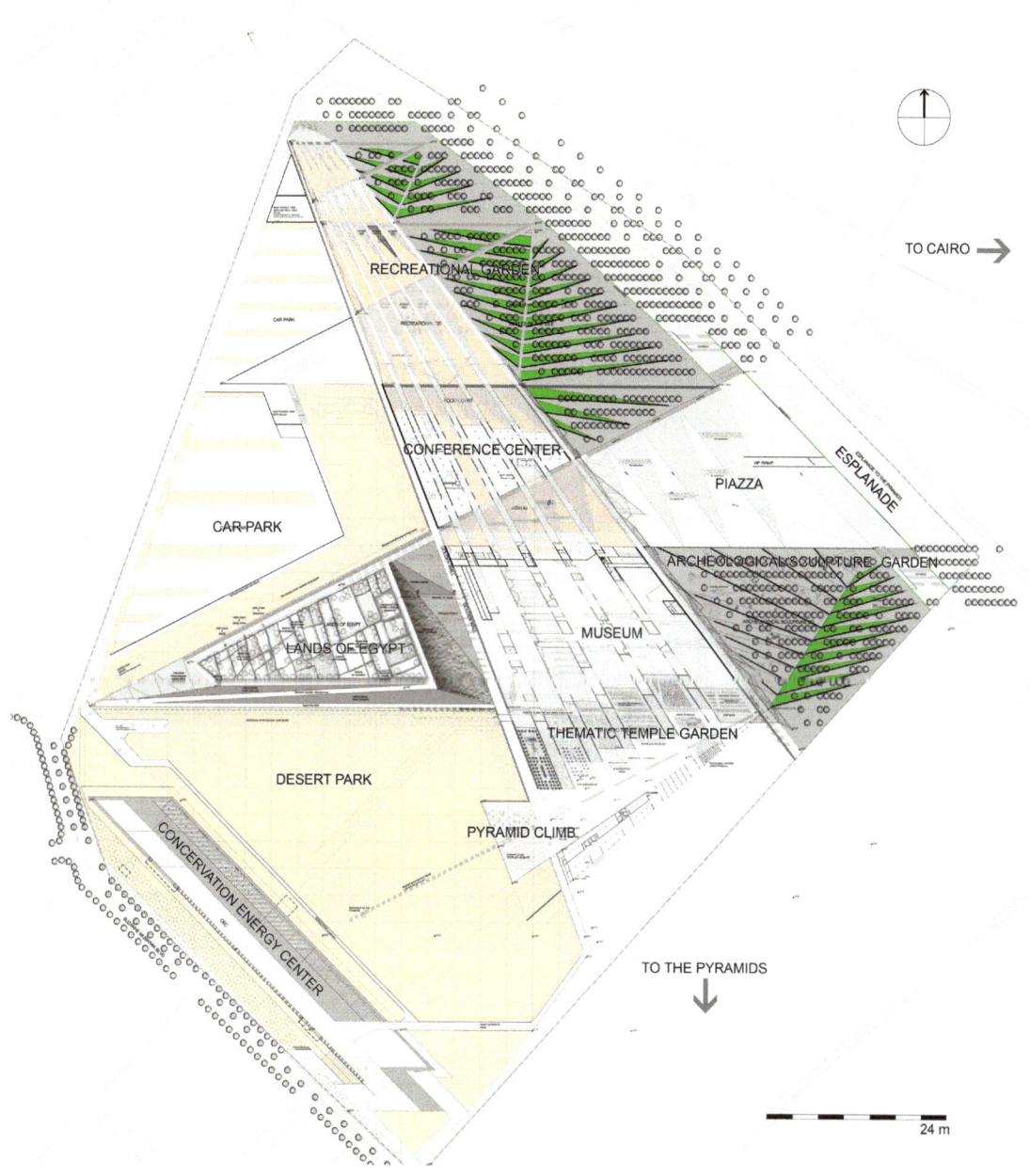

16.6
Grand Egyptian Museum, Cairo, Egypt. Collaboratively working in radical accuracy, across multiple scales. Delicately rendered CAD plan with grey hatching.

the bike as the main character, with the new park as a beating green heart within a blood-coloured New York Harbour (see Figure 16.10).

Abstracting

Abstracting the context can be used to reposition the shape as a semi-identifiable motif back to the viewer to further digest and seek inner meaning. In the design of the Pergola at Máximapark in the Netherlands, the exploration, testing, and prototyping options were an evolutionary process to create a constructable yet adaptable module. Inspired by the repetition and modulation of M.C. Escher, the design evokes the illusion of infinite transformations within the framework, manifesting the sublime (see Figure 16.11a and 16.11b).

16.7
Toronto Waterfront, Lake Ontario, Canada. Iconic perspective capturing the spirit of a Canadian winter pasttime, skating on Lake Ontario.

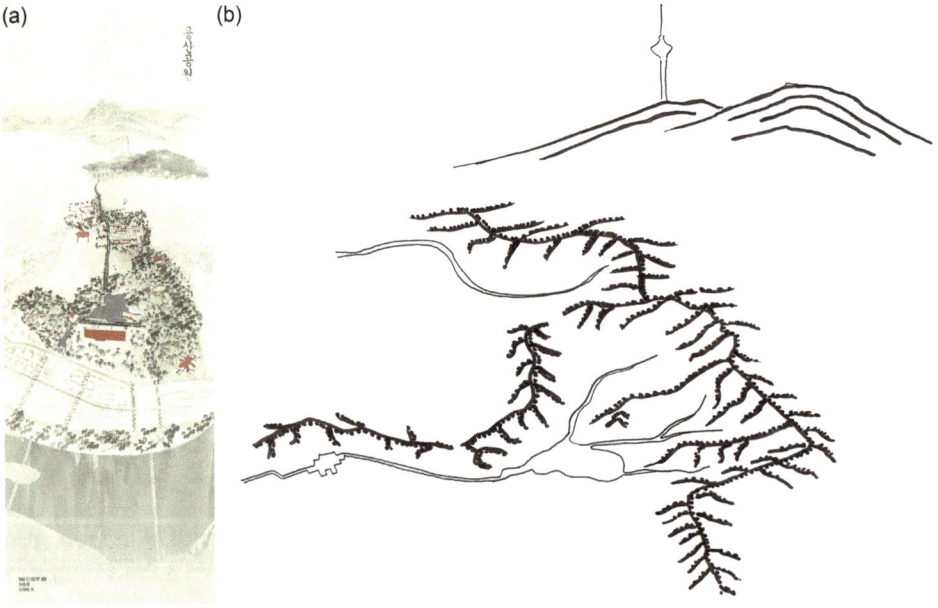

16.8
(a) and (b) Yongsan Park, Seoul, South Korea. Expressing the rich layers of designing thinking in the master plan by combining computer-generated images with the Korean traditions of landscape painting.

Freedom from an Innocent Landscape

16.9
(a) and (b) Yongsan Park, Seoul, South Korea. Creating an image where the viewer feels like they want to be there; dynamic, liveable, and realistic spaces interpreted with techniques of 'aquarelle', pencil colours, papyrus, and silhouettes of light.

Similarly, to express a composition for a series of murals representing the universal nature of agriculture, cultivation, wellness, and cuisine, the murals of the water walls in the Edible Garden at Houston Botanic Garden combined the processes of 2D and 3D manipulation, pattern-making, and colour combinations of real tiles to achieve the final result (see Figure 16.12a and 16.2b).

The Hand of the Painter

Despite the benefits of today's technology, where anyone can generate a landscape at the click of a button, West 8 has continued to use the above techniques to both the benefit of the design and the strength of the visual communication. West 8 utilizes the hand of the designer to provide the final touch.

Landscape and public space are not viewed as a mere collection of objects, cladding, and furniture placed at random within a scene—but as a provocative, composed scene that establishes itself at the cutting edge of place-making and design of the urban realm. Pushing beyond the expected images of an innocent landscape, West 8 abstract, distort, and place iconography for the viewer as cultural touchstones to create images filled with life.

16.10
Governors Island, New York, USA. Re-positioning the forgotten Governors Island in the mind of the average New Yorker; poster showing the bike as the main character, with the new park as a green, beating green heart within a blood-coloured New York Harbor.

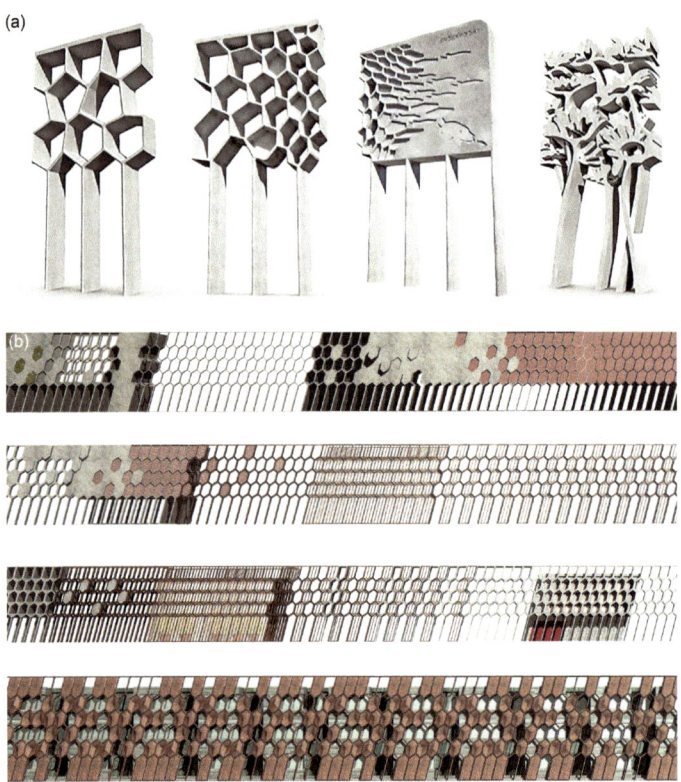

16.11
(a) and (b) Pergola of Maximapark, Utrecht, the Netherlands. 3D digital models and abstract renderings of adaptable modules, the ability to rapidly test in real-time the output of minor design changes across a 3.5-km long pergola.

16.12
(a) and (b) Houston Botanic Garden, Texas, USA. Abstraction and transformation in real-time; Creating at the Culinary Garden included a process of 2D and 3D manipulation, pattern-making, and working with the specifications of real, hand-made tiles to test the design of the three, jade-coloured waterwalls in real-time.

The field of landscape architecture all too often ignores the contrasts inherent of today, avoiding the crude realities that offer opportunities for conflict or (self-)reflection. Should landscape be limited to representations that have lost their identity of place? Why represent an unrealistic product of songbirds, springtime flowers, butterflies, and children with balloons?

The systemic, laborious approach to composition has always been part of the creative design process of West 8 and it is not long-gone as a relic of a past era. Instead, the designer should introduce representations and impressions that capture the concept and the narrative, avoid shallow illusions of mass culture, and free yourself from the preoccupations of the romantic paradigm of the 'innocent.'

17 Evolving Representation, Physical and Digital at Hargreaves Jones Landscape Architecture

Matt Perotto

For the past 30 years, Hargreaves Jones Landscape Architecture (formerly Hargreaves Associates) has been designing and implementing world-class landscapes across the globe. Over this time, the firm has received over 100 awards for its transformative projects and has developed a portfolio of published and exhibited works nationally and internationally. Projects range from large-scale, environmentally complex brownfield sites, waterfronts, and campuses to small-scale urban plazas and gardens. In many instances, the projects commence with urban or open space master plans that transition into site design and implementation.

After receiving the invitation to contribute to this publication, we began by sifting through the firm's project archives, creating a catalogue of various representation techniques over the firm's history. Through this process, one thing became abundantly clear, while many of the tools, techniques, and mediums used to communicate ideas have evolved, the underlying intended dialogues within each project phase remain essentially the same. This reflection became the launching point for the structure of this chapter. Rather than presenting the firm's projects and associated visual representation techniques from a perspective of temporal change over 30 years, this chapter aims to discuss how these techniques have evolved within each project phase in service of the design process.

Site Acquaintance

Projects always begin with the process of learning about new contexts. Every site is different, forming an intimate relationship with the place, and its context is a necessary jumping-off point for the firm's design process. This always includes an in-depth inventory of all available secondary resources, both spatial and non-spatial (shapefile data, surveys, aerial imagery, historical information, demographics, etc.) as well as primary photographs via site visits. Site photography plays a critical role at this stage—the act of which forms an accurate visual representation of the site's existing conditions experienced first-hand.

Boxes of photo negatives sitting in any one of the Hargreaves Jones Landscape Architecture offices will attest to the volume of photos captured at both project onset and completion (of course, since the advancement of digital photography, photos no longer require physical storage). This primary and secondary research is then combined to create visual site-orientation packages that are shared amongst the design team—at one point, selected physical photographs and other resources were tacked to pin-up boards in the office, whereas the contemporary workflow requires printing of the digital package for the same purpose. Key moments that emerge from the site visit or through investigation of the orientation package are flagged for further in-depth analysis, which often includes the drafting of sections and elevations to illustrate relationships between adjacent elements. Sometimes, the act of discovering and distilling site constraints into a simplified graphic provides a key motivator for a large-scale conceptual design gesture. This process was the case for the Queen Elizabeth Olympic Park, London, UK.

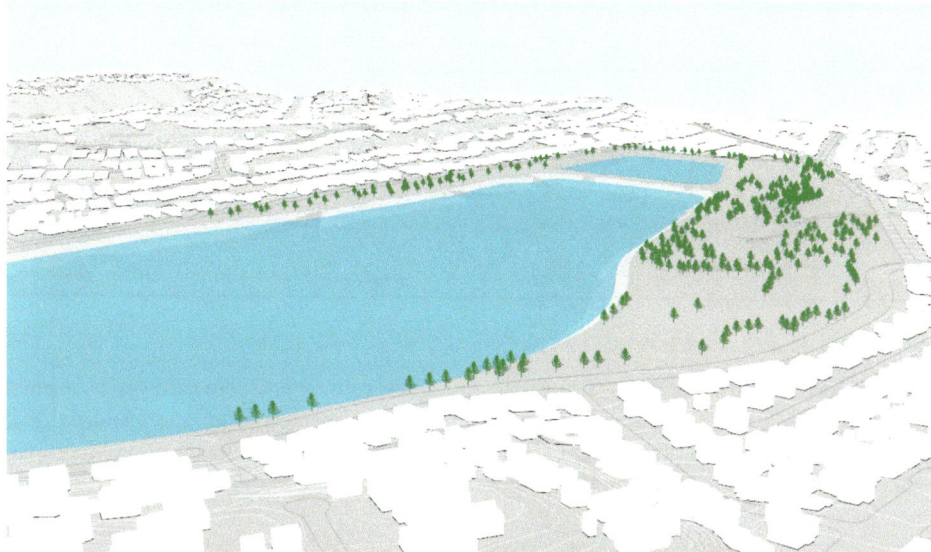

17.1
Site analysis model views. A digital model was created during the inventory and analysis phase of the Silver Lake Reservoir Complex Master Plan, Los Angeles, which was used to generate perspective views and design drawings of the site and context for further analysis and design conceptualization.

With the proliferation of available digital data in most contexts (surveys, shapefile data, aerial imagery, etc.) and intuitive 3D modelling software, it has become common practice for the design team to create digital 3D models which allow for further detailed investigations. Models will include anything from the geomorphological and ecological (complex topographic conditions, hydrological elements, important tree and planting stock) to the site's cultural inventory (buildings, circulation, program areas, etc.). However, the resolution of details embedded in the digital model remains only enough to generate visuals effective for further site analysis—including aerial and eye-level perspective views, section cuts, and elevations (Figure 17.1). Firm-wide standardized colours are applied to model surfaces to add a layer of efficiency in modelling and analysis, and 3D-scale representations of 2D people are inserted to allow for an effective understanding of scale. These site-orientation packages are often presented to the client to communicate key findings and solicit feedback. By the end of these meetings, the design team has identified critical aspects of the site that will guide the entry into design phases.

Conceptual Design

Sketching has and will continue to play the most crucial first step for design exploration. Spatial ideas are drawn above the produced inventory plans, sections, and perspectives, and non-spatial diagrams help facilitate the internal sharing of ideas. The next step involves representing these initial ideas in three-dimensional space for review, discussion, and revision.

In early projects like Candlestick Park, San Francisco, CA, and Byxbee Park, Palo Alto, CA, the firm utilized a physical four-foot square sandbox constructed in the office to explore design ideas. Sand proved to be an effective medium for fast-paced model iterations, allowing several individuals to push the sand around simultaneously with the immediate visual reflection of each iteration (Figure 17.2). Uniquely, it also allowed for tangible, physical representations of a natural angle of repose; slopes that would be too steep to implement in construction were also too steep to model as the sand eroded down on itself. On the other hand, the preservation of 'modelled' ideas proved to be a challenge. Photography was required to 'bake' a design idea for future reference and replication once the sand model was altered. Furthermore, while the sand model allowed for effective collaboration amongst the internal design team, it did not allow transportation off-site to present formalized ideas with the client group.

17.2
Sand to refined model: Candlestick Park, CA. At the Hargreaves Jones San Francisco office, a physical sand model was used to study a diverse array of landform massing and geometries for the Candlestick Park, San Francisco project. The design team crafted the final forms generated using sand to model the topography of the landscape. The landforms are evident in the more permanent presentation models and the eventual site intervention.

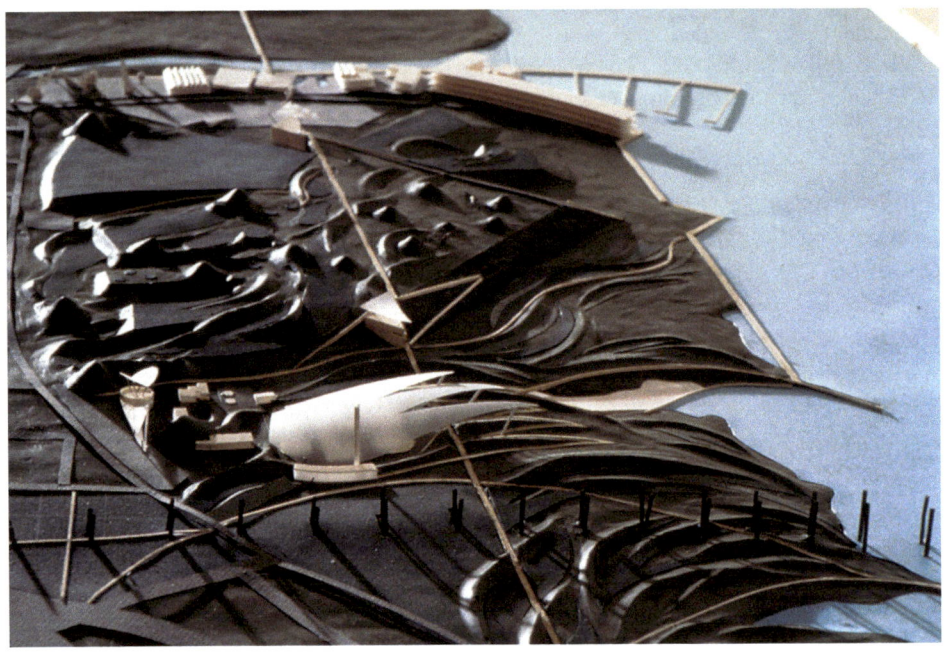

17.3
Clay model: Lisbon. Clay modelling was a critical component of the Hargreaves Jones winning competition proposal for the 1998 Lisbon Exposition, Portugal. The physical model was developed during early design exploration, and the model allowed the design team to carefully study the manipulation of the brownfield landscape in relation to circulation and program.

To resolve these issues of permanence, clay models began to replace the use of the office sandbox. As a medium, clay takes more time to model than sand but allows for much more versatility in formal topographical expressions and more detail and accuracy in volumetric studies. Clay also provides a sense of permanence since the model can be moved, rotated, and tilted for analysis and transportation purposes. As a result, clay models were used as the primary source of ideation in many of the firm's notable projects, including the Commons and Campus Green at the University of Cincinnati, OH, The Clinton Presidential Center in Little Rock, AR, Louisville Waterfront Park, KY, and the 1998 Lisbon Exposition in Portugal (Figure 17.3).

Over the past 30 years, Hargreaves Jones Landscape Architecture has seen a fundamental shift in the scope, structure, and delivery of landscape projects which have required the collaboration process to evolve. Design teams are no longer composed of landscape architects with one or two sub-consultants with information exchanges in a silo. Instead, contemporary projects often require broad interdisciplinary teams, with the teams frequently sharing information. This has guided the firm (and the discipline in general) to explore alternative, digital methods of 3D modelling that enable the flow and exchange of information without losing the investigative nature of conceptualization achieved through physical modelling. Digital 3D modelling allows for testing a multitude of design ideas within one interface at a real-world 1:1 scale. Furthermore, by modelling the existing conditions model generated during the inventory process, site-specific variables can now be factored in the iteration process. The versatility in camera positioning and drawing outputs allows for a much more timely and flexible evaluation of the design alternatives. Digital models also allow for efficient information exchange amongst large design teams. For example, rather than only sharing 2D drafted drawings for ongoing coordination processes, 3D models are swapped with other consultants, which then get linked directly into our working model for continued spatial reference.

Client & Community Engagement

All projects require coordination and collaboration with the client group(s). When projects have public clients, the design team works with the public through community engagement processes to garner critical input and feedback. To help facilitate these dialogues, the design team produces diagrams and illustrative drawings (plans, sections, elevations, and perspectives) to visually represent ideas to date—but not everyone can understand design drawings.

Physical models for community engagement are usually more polished than the initial study models previously discussed. For example, during the community engagement processes for the first phase of the Core to Shore Master Plan in downtown Oklahoma City, OK, a physical model was developed for each of the three conceptual designs for Upper Park. The iteration models were critical in helping members of the community visually analyse their 'likes' and 'dislikes' of each concept. In later meetings, the feedback received from the community translated into a consolidated design represented in one final physical model, a design that has now been implemented at the newly named Scissortail Park. The level of permanence desired is a critical factor driving the material selection for presentation models. Clay massing, cork and millboard layering, wood, and foam are all options to form the model base. Clay offers a malleable medium with an implication that the output isn't precious, and if the model will continue to be tooled and retooled, then clay becomes a clear choice (Figure 17.4a). On the other hand, foam and cardstock can represent a more refined level of design, but paper cannot be repaired once it's cut. Once the base is developed, additional elements for scale and relation are added like trees, cars, people, etc (Figure 17.4b). In the same gesture as diagramming, coloured elements are often used to highlight important moments in the site design to allow for the most effective communication of the design intent.

Physical models always augment illustrative drawings and visualizations. Hand drawings and collages formed the basis of graphic representation earlier in the firm's project portfolio, whereas digital drawings are now directly generated from digital 3D models. Methods and tools for producing digital graphic media continue to evolve, and the firm is constantly exploring new software for the effective production of visuals in an efficient timeframe. Depending on the intended audience, illustrative drawings and views from the model may be entirely created via image editing, or they may be pushed through a rendering engine to include textures and elements such as vegetation. This process was followed for the Silver Lake Reservoir Master Plan, Los Angeles, CA (Figure 17.5).

The firm is interested in new tools and is continually advancing how to engage the community. In pursuit of new methods of public design proposal interaction, at one of the community meetings for Northside Park in Hunters Point, San Francisco, CA, an illustrative site plan and perspective visualization were printed on large format banners. The site plan was scaled at 1 in = 10 ft and was placed on the floor in the centre of the room, where attendees were invited to physically walk through the park on guided tours by the design team (Figure 17.6). Similarly, the perspective visualization was hung on an exterior wall of the building where visitors were encouraged to enjoy a barbecued meal and have their photo taken with the banner in the background as if they were having a picnic at the park.

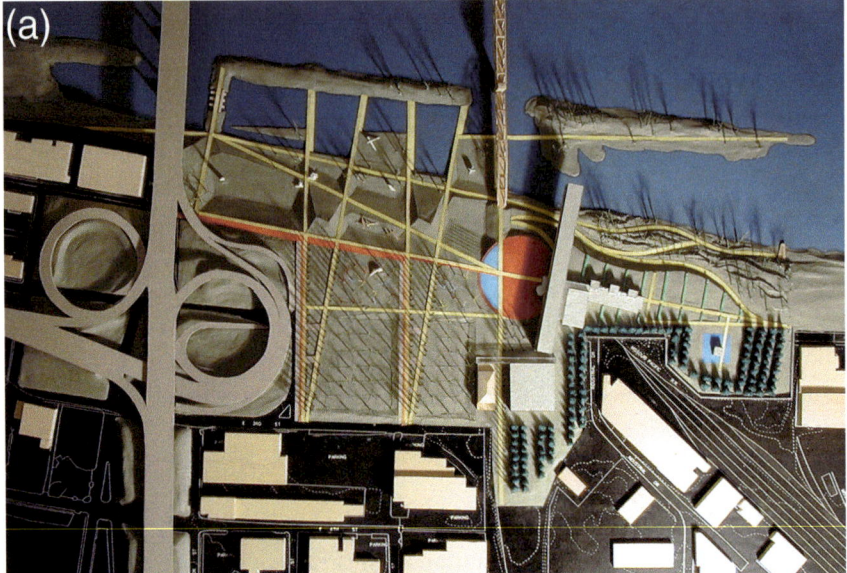

17.4
(a) Physical presentation models (working model). This model was created of the design of the William J. Clinton Presidential Center, Little Rock, AR. This model was created as a tool to refine relationships between topography, hydrology, circulation, and program. Coloured construction paper was superimposed over the clay base to highlight specific areas on the site and to allow for more discussion and analysis. (b) Physical presentation models (refined model). This presentation model was developed for Louisville Waterfront Park, KY. It was an important aspect to communicate the design intent to the clients and the public. Hargreaves Jones continues to use clay models as part of their visual communications toolkit.

Refining the Design

Once a design direction is determined and the project continues forward, additional types of visual representation are utilized to progress design detailing. Physical and digital models continue to be an important tool in refining spatial forms and detailing relationships between adjacent site elements. During the design development of Queen Elizabeth Olympic Park, London, UK, a combination of clay, cardstock, and foam was used to model refined iterations of smooth, sinuous curves and precise landform planes that had to preserve a close balance of cut and fill. An existing condition clay model was utilized as a baseline volume of available clay (earth) for manipulation to accomplish this. This physical, reshaping process allowed for explorations of cut and fill volume exchange across the entire

17.5
Lumion generated renderings. Renderings developed for the Silver Lake Reservoir Master Plan, Los Angeles, CA, were generated using a rendering engine to computationally render material textures, plants, sky, sun, and shadows, before being refined in a photo editor to add people, animals, and final colour touch-ups.

17.6
Walk-on site plan. During the community meeting processes for Northside Park in Hunters Point, San Francisco, CA, guided tours of the proposed site design allowed members of the community to engage with the design proposal in a more purposeful manner than simply looking at drawings on presentation displays. For example, parents were able to walk their children down the main pathways and point out the playgrounds, basketball courts, and the shoreline. This visual helped to generate immediate emotional connections and a sense of excitement for meeting visitors.

17.7
Physical refinement model. A variety of physical models were created during the Queen Elizabeth Olympic Park, London, UK, design process which merged into this model. The clay base was used to help balance a rough estimation of cut-fill throughout the north section of the Olympic grounds. Pins were then added for tree massing, and cardstock was laid above to communicate different ecological types and the threading of walkways within which navigated the elevation delta between the event venues and the River Lea.

site, while simultaneously studying and finalizing the actual landform gestures (Figure 17.7). While this cut-fill exercise only allowed for rough estimation, the process ensured that volumes were considered before the transposition of physical to digital for computational cut-fill analysis, ultimately saving time and money in the design process.

Once physical models were finalized (pre- digital modelling software), pins were inserted into the clay to take to-scale survey elevations. These were then imported into a 2D-drafting program, and contours were interpolated to produce a grading plan. This practice was common practice in projects like the Clinton Presidential Center, Little Rock, AR, and Louisville Waterfront Park, KY. In the firm's current workflow, a digital model is maintained in tandem with the physical model. If the physical is utilized to generate an idea for further detailed study, it is explored in the digital realm. Landform surfaces and other massings within the digital model are then used to generate contours for grading plans. This technique can be especially beneficial in translating complex forms such as circumferential slopes, which blend between gradients into a grading plan, as was the case of the landform hiding the central utility plant at Google's Flagship Headquarters Charleston East in Mountain View, CA.

As design refinement progresses, aspects of the design are considered in more detail via the study of alternatives. For example, program areas are identified at the concept and schematic design stages, which indicate applicable ground plane materials. Pavers may have been identified, but we now study the shape, size, pattern, colour, etc., to arrive at finalized detail and products. Studies never happen in an isolated vacuum and always consider the adjacent site elements and the larger comprehensive design concept. Digitally modelling these iterations within the context of the site model is critical to the process, and iteration view packages produced allow for the visual analysis of options from the same vantage points.

One of the new tools that the firm has recently begun employing is parametric computation. Common in the architectural office, parametric computation allows for creating visual algorithms to generate digital models. Whereas traditional digital modelling calls for the user to model an element, parametrics requires the user to model the process—this is very similar to how we annotate typical construction details. For example, at the Federal Building Plaza in San Francisco, CA, a script was developed to model a fence. This was an important feature of the plaza, which allowed for testing the form, size, and spacing of pickets and horizontal rails, considering the pre-defined code-related rules, the structural requirements, and the overall aesthetic impact of the intervention. Once the script was developed, the design team was able to test different inputs and visually analyse the modelled output live during meetings, dramatically increasing the speed, efficiency, and collaborative nature of the detail refinement decision-making process. In this manner, visual representation is no longer predetermined prior to meetings with static imagery or even dynamic video, but now becomes interactive for on-the-fly iteration.

Getting It Built!

The intent of the construction documentation process has largely remained the same over the past 30 years. However, the tools used to finalize the design and prepare the construction drawings continue to evolve. In the pre-digital drafting days, construction drawings were drafted by hand, and a library of physically drawn details was stored and maintained in flat filing cabinets. More recently, digitally drafted construction detail standards are saved in digital libraries for future project reference and revised based on lessons learned, in much the same way as before.

This process is beginning to change with the firm's transition to the new world of Building Information Modelling (BIM) on projects such as 'The Eight,' a mixed-use office tower and public plaza in Bellevue, WA. Using BIM, one central model is developed by each discipline, starting in early design phases. The modelled content has information embedded within each material or family (a pre-defined element) which can be used to analyze the design for an ever-increasing range of quantifiable metrics. As projects progress, models are shared amongst the design team to coordinate details and identify and resolve clashes between scopes (Figure 17.8). More detail is embedded into the model (defined in a project-specific BIM execution plan). In the construction documentation phase, plans, sections, enlargements, and detail views are generated directly from the digital model. Views are then added to sheet layouts and after minor manual drafting overlays and annotations, the drawing package is complete. All drawings pull information from one model, and if issues arise during construction that requires design adjustment, the revisions are made to the model, and the drawings are automatically updated and ready for re-issue.

Complex digital models such as these provide the design team with a resource with which to create additional presentation graphics to tangibly communicate complex interfaces between architecture, engineering, and landscape scopes for informative purposes. This remains an important aspect of the design and implementation process to ensure the public, client, and contractors remain informed of the design intent (Figure 17.9).

The Landscape, Post-Occupancy

After implementation, the firm always makes a point to end where we started, that is, back to photography. Photos help the office to validate the final finished outcome of the design intent through direct visual representation of the landscape. This self-reflection process is a critical aspect to a successful practice, analyzing how the graphic representation of design ideas is now manifested in physical space, if program areas are functioning, ecologies are establishing and flourishing, and users are engaging with the space as intended, or if serendipitous outcomes are observed (Figure 17.10).

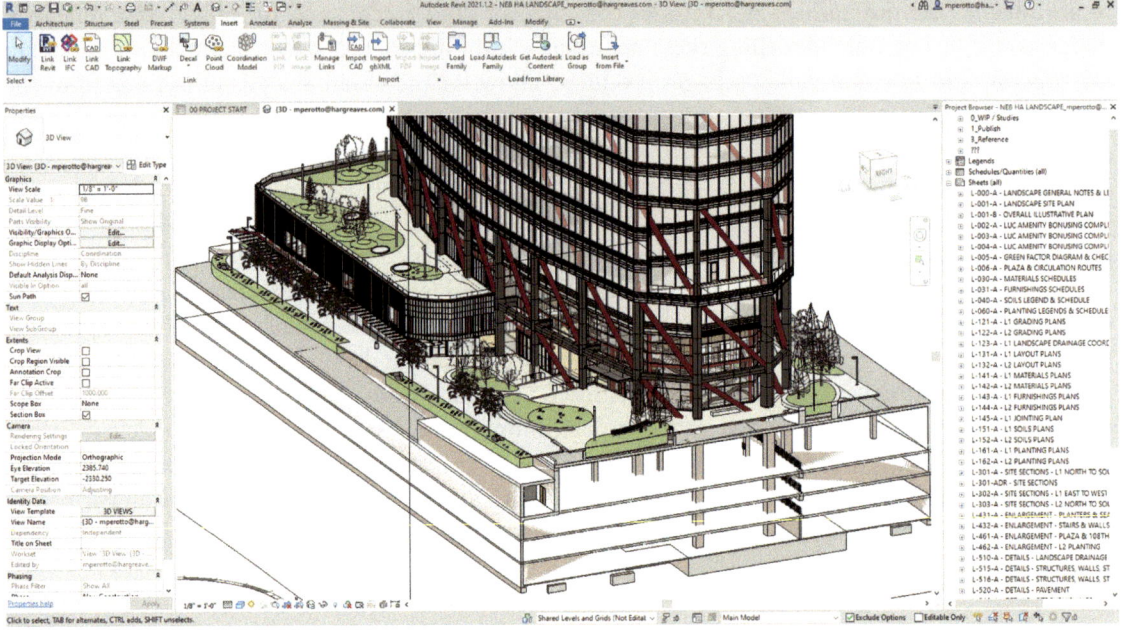

17.8
Revit model screenshot. Revit was used throughout the design process at 'the Eight,' Bellevue, WA, a mixed-use commercial development, where plaza and streetscape landscapes negotiate 4.5% existing slopes along the site perimeter and all sit on structure. A rooftop terrace above restaurant and retail space also required careful coordination with the project team.

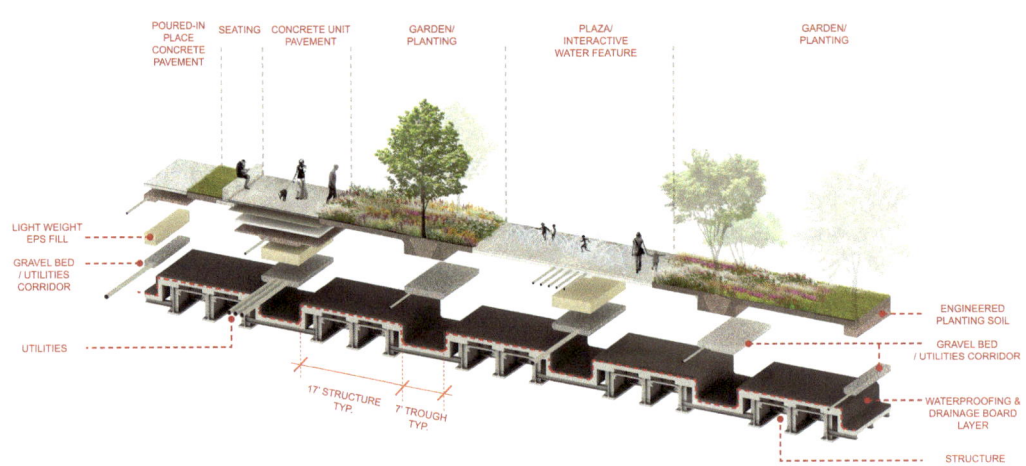

17.9
Complex interfaces. The Penn's Landing, Philadelphia, PA, project involves the creation of an on-structure cap over the I-95 and Columbus Boulevard along a section of Philadelphia's riverfront. Detailed coordination between the project team, client, and regulator agencies has been required throughout the Feasibility Plan and into the detailed design process.

In addition, project certification in sustainable design-rating systems, such as the USGBC's LEED and SITES systems, is becoming more common in Hargreaves Jones' projects. These rating systems require an analysis of sustainable design principles and metrics during the design process and post-implementation to ensure projects make a meaningful contribution to their sociocultural, economic, and ecological contexts. The development of tangible graphics provides the means for discussion and analysis.

17.10
Post-occupancy, Crissy Field, CA. The 100-acre conversion of the U.S. military installation at the Presidio, San Francisco, into a national park, Crissy Field. This was completed in 2001. Since then, it has become an integral part of the sociocultural fabric of the Bay Area and plays an important role in the ongoing restoration and rehabilitation of the natural landscape of wetlands, dune field, and beach along the north Shores of the City.

Over the past 30 years of Hargreaves Jones' practice, much has changed concerning how people engage with landscapes (and what people want out of their public space); associated clients' desires (both in the public and private realms); design team composition; construction methodologies; and how we collaborate with the constituents of each. Projects are becoming much more complex, and architecture and landscape continue to meld in archetypal ways. To stay effective throughout these changes, the contemporary landscape architecture firm's approach to visual representation and communication must evolve and stay on the leading edge of these changes—increased complexity, broader collaboration, and maintaining an overall efficiency through the process.

This gesture wholeheartedly is embraced at Hargreaves Jones. Interestingly enough, while the firm continues to make a conscious effort to explore new methodologies for visual representation, much of the same techniques employed in the earlier work (physical model making, for example) still play an important role in the contemporary practice in conjunction with an ever-growing repertoire.

18 Drawing *in* Perspective

David Malda

During a visit to Gustafson Guthrie Nichol (GGN Ltd) in Seattle, it is not uncommon to see a team of designers kneeling around a drawing placed on a table with their eyes just above the level of the paper. Rather than a traditional pin up with drawings on a wall and the team lined up in response, this simple shift in position is a key legacy of Kathryn Gustafson's influence on the firm she cofounded 20 years ago. Long recognized for her use of clay models in design, Gustafson's approach to drawing challenges traditional relationships among the body, representation, and making in landscape architecture. While the link between sculpted clay models and sculptural landforms has been featured of her designs over the past 30 years,[1] the influence of modelling offers broader lessons for drawing and hybrid forms of making. These connect a conceptually driven approach influenced by modern composition with the picturesque emphasis on movement and experience through time. While concept and narrative are central to Gustafson's work, it is the focus on the acts of making that will be the focus of this essay.

Drawing to See

Drawing is central to the practice of landscape architecture. Designers draw to articulate ideas through concept sketches and diagrams. These representations externalize thinking, making it accessible for self-critique and collaboration. One draws in order to see an idea and share it with others. From there, the drawing takes on its own life, eliciting response through new drawings that refine, clarify, or intentionally contradict. The design process culminates in yet more drawings that instruct the many hands responsible for physical construction. These too are abstractions that rely on well-established conventions to ensure the translation from two-dimensional images to built form achieves the desired outcome.

Drawing's central role in developing both the content of a design and the means by which that content is translated into built form highlights a tension in the design process. Before drawings can explain how to build, they first need to explore what to build. This relationship often connects to two primary conventions of representation in Western traditions—the perspective that explores a design as it might be experienced once built and parallel projection that is central to conventions for guiding construction.

The merits of these two drawing conventions have long been debated within architecture.[2] How one sees and communicates ideas through representation directly impacts the content of design. Parallel projection is the key to the plans, sections, and elevations that have long served as the primary drawings for constructing landscapes and buildings. They provide predictable information about scale and the relationships among the various elements of a design. The plan or elevation allows one to work the entirety of the design, simultaneously considering the relationships among various elements. By being parallel in projection (or the relationship between the picture plane and the viewer), there is no convergence so no one point of view is given priority over another. The viewer ceases to be a body

DOI: 10.4324/9781003183402-18

in space and is given equal access to the entirety of the design as represented. Its measurability also means it's good for guiding construction.

As a design tool, this total nature of the plan is often criticized for failing to translate to the human understanding. Patterns and geometries that might appear compelling as two-dimensional compositions don't necessarily translate to compelling spatial experience. A familiar chorus in countless design reviews is the reminder that "no one will ever experience the project in plan or elevation."

The perspective, unlike parallel projection, intends to privilege human experience from a specific point of view. Some begin as an imagined scene that will later need to be evaluated for the spatial relationships implied. Others rely on Renaissance traditions to geometrically construct a view from established plan and elevational information. Regardless of approach, these conventions rely on translation from one drawing type to the other to connect a specific moment of experience with the comprehensive description of the project. The two are linked by a common understanding of space within which the project exists.[3]

While the perspective is often treated as closer to human experience than the parallel projection, it has also been long acknowledged to have limitations. Beyond the nuances of optics of the eye, the fixed viewpoint of the perspective necessitates the drawer to choose a predetermined viewpoint and from there remains static. Much of drawing and painting over the past century explored challenges to the fixed nature of representational painting and drawing, building on growing science that the human eye is in constant motion while seeing and that movement is closely tied to our experience of a place.

The multiplicity and fragmentation in painting and collage introduced as alternatives to traditions of static realism rarely challenged a consistent relationship between the viewer's body and the image. The plane of the picture and the viewer was aligned and constant. While this convention is fundamental to the constructed perspective, it has generally been less emphasized as foundational for parallel projection as well. The multiple viewpoints permitted by parallel projection mean that there is no single vanishing point around which the drawing is constructed, but still assuming that viewing the drawing the orientation of the eye to the picture plane is consistent, even if the viewers can move side to side or up and down without affect. Rotating the drawing in relation to the viewer introduces distortions breaking the relationship between the reading of the image and the information contained.

Remove the drawing from the wall. Place it on the table. Now lower your eye to the table and look across the paper. Draw.

This simple move fundamentally transforms the drawing. By breaking the fixed relationship between one's body and the picture plane, the drawing ceases to be confined to the conventions of two-dimensional representation. It becomes an object in space to be engaged as would be expected of a sculpture or model. If the drawing began as a plan, moving the eye down to the level of the table and looking across the paper will shift the experience to that of a perspective. Parallel lines converge. Elements closer to your eye appear larger than those further away. Moving back above the drawing and looking down returns the relationship to that of parallel projection. The plan reads as a plan once again.

Approaching drawing through the spatial and experiential qualities of sculpture is a lesson Kathryn Gustafson passes on to designers at the landscape architectural practices she cofounded in Seattle and London. Gustafson has made sculpture, in the form of clay landform models, central to her design process since she began practicing landscape architecture in the early 1990s. Unlike many architectural models developed as a presentation tool after the design is completed, Gustafson uses the process of modelling in clay to move from early concept drawings to refined design. Typically, initial plan geometry is inscribed onto a slab of clay and then carved down to achieve the desired topography. Throughout the model's development Gustafson and her collaborators are constantly in motion—testing specific views and spatial relationships. The composition is refined until the many aspects of the intended experience are unified through the model's surface. Eventually, a copy of the model is cast in plaster then photographed or digitally scanned to become the basis of technical drawings for continued refinement. Even then, the model lives on as a reference against ongoing development. Though typically presented in pristine white relief, there is often a second plaster casting that continues to be drawn on and manipulated as the design progresses.

While only a portion of the projects at Gustafson Guthrie Nichol are developed in clay model, the approach described above extends to a culture of drawing and making throughout the studio. Bringing drawings onto the table and working from multiple vantage points blurs the lines between drawing and model and highlights the importance of connecting form with experience. As Gustafson likes to point out, "You can't cheat in model."[4] Unlike a perspective drawing that can present a compelling idea with an uncertain relationship to the conditions and constraints of the project, shifting between plan and perspective in a single model or drawing ensures that both interpretations are tied to a comprehensive design.

Gustafson continues to be fascinated by the relationship between form and experience. In lectures, she often shares examples of paintings by Kandinsky and Malevich, first as one would typically see them represented in a book or on a gallery wall, then when viewed from a low angle across the work. The lesson is powerful. Not only is the experience of such works drastically transformed through the different points of view, but the nature of the forms provides a rich and dynamic experience as one "moves" through them. Such works often serve as inspiration in Gustafson's designs, and her work has been described in terms of modern land sculpture.

While some designs begin as more traditional compositional studies in plan, Gustafson frequently shifts to drawing "in" perspective with her eye at the level of the table. From this position, it is easy to see the influence of working in clay—the pencil is pulled across the drawing as one might pass a tool through clay. The drawing is worked, viewpoint shifted, and refined. Significant in this approach is that, unlike traditional painting and drawing, where the eye and the hand are in parallel planes, Gustafson splits the plane of the eye away from that of the hand, just as the body would stand perpendicular to the surface of the ground. The eye is vertical, and the hand is working horizontally, allowing both conditions to be developed simultaneously and leveraging the benefits of what would typically be understood as a distortion of the view. Unlike anamorphic drawing which is also based on manipulating the distortion created when the eye and drawing plane are misaligned (Hans Holbein's skull in *The French Ambassadors* being a well-known example[5]), drawing from a shifted viewpoint in Gustafson's work is never intended to recreate a singular optical trick from an alternative viewpoint. Rather, it is by drawing from many viewpoints that the overall composition of the design takes shape.

Working fluidly among multiple viewpoints supports another important aspect of Gustafson's designs. Building on picturesque traditions, her work is intended to be appreciated in motion.[6] Though compositionally unified, these landscapes are never fully comprehended from any one viewpoint. The shape of the land is designed to lead one through a space, guiding movement from one moment to the next as the experience unfolds. Working in model and model-drawing hybrids offers a direct link between the movement of one's body in the process of making and the intended effects of the built work. Shifting viewpoints around the drawing simulates the experience of movement with immediate feedback. Unlike static views or even predetermined animations, there is no limit to the range of motion.

Drawing-model hybrids extend beyond the plan drawing techniques described here. Going from model to plan and back again is a frequent cycle as three-dimensional information is incorporated. This method comes in the form of spot elevations and slopes added to drawing that help the team see the topographic relationships as the plan evolve. Detailed grading early in the process (along with accompanying section studies) has been an essential component of Gustafson's work as, like the clay model, it ensures that the constraints of reality all add up to a design that is achievable. While for many these annotations appear more cryptic than informative, Gustafson has learned to mentally translate this information into three-dimensional form while working. The immediacy of this approach allows a great deal of precision in the designs.

At times, early concept drawings become models through a process of cutting and folding that allows abstract three-dimensional relationships to develop out of the original sketches. Within the practice, this kind of direct translation from one drawing to the next is considered an essential part of maintaining the subtle and intuitive forms of making throughout design. Sketches are copied to develop multiple iterations. Scale figures such as people (which Gustafson also adds to plan drawings to be viewed in perspective) return these abstract conceptual models to an immediate relationship with the body. Once developed these studies are often scanned or photographed to inform the next level of development.

Digital modelling also plays a role in this process. In many respects, an application like Rhinoceros® (Rhino3D® digital modelling software) is designed to do much of what has been described above. It allows one to view a project through plan, elevation, and perspectives simultaneously in adjacent viewports. Because these views are constructed from the same digital information, updating an element in any viewport updates all viewports. While digital modelling is a key component of developing design at Gustafson Guthrie Nichol, it has yet to replace physical drawing and making in the design process. This is partially due to the subtle and intuitive capacity of the hand in motion, but it is also related to the limitations of digital modelling applications with regard to the idea of drawing in perspective. While developing a physical drawing, the hand and the eye are able to simultaneously work in independent planes. The upright eye sees the image as perspective but the hand works in plan. Digital modelling applications allow one to work in plan and perspective simultaneously, but the disconnect between the independent viewports for these two projections means that the physical connection to the hand cannot offer the same kind of complementary feedback as it would when working on a physical drawing. For this reason, digital modelling applications tend to be used to augment and refine designs developed by physical means. Two-dimensional and three-dimensional information is scanned and further developed in applications like Rhinoceros®.

Though well established in techniques of pencil and clay, Kathryn Gustafson continues to advocate new means of improving the connection between experience and design. Increasingly, project teams utilize virtual reality and physical printing to study refinements and ensure that the early ideas translate through to physical construction. Like the blurring of boundaries between drawing and sculpture that has been central to Gustafson's process for decades past, she continues to explore new opportunities to connect the process of design with human experience in ways that resonate with our time. Refer to Figures 18.1–18.6, which showcase GGN's design process using sketching, paper study models, clay-modelling, to 3D printing.

18.1

Kathryn Gustafson with Neil Porter of Gustafson Porter + Bowman during the development of the Diana, Princess of Wales Memorial Fountain.

18.2
Early concept drawing, Hazelwood Green, by GGN.

18.3
Paper study model with scale figure, Hazelwood Green, by GGN.

Drawing *in* Perspective 159

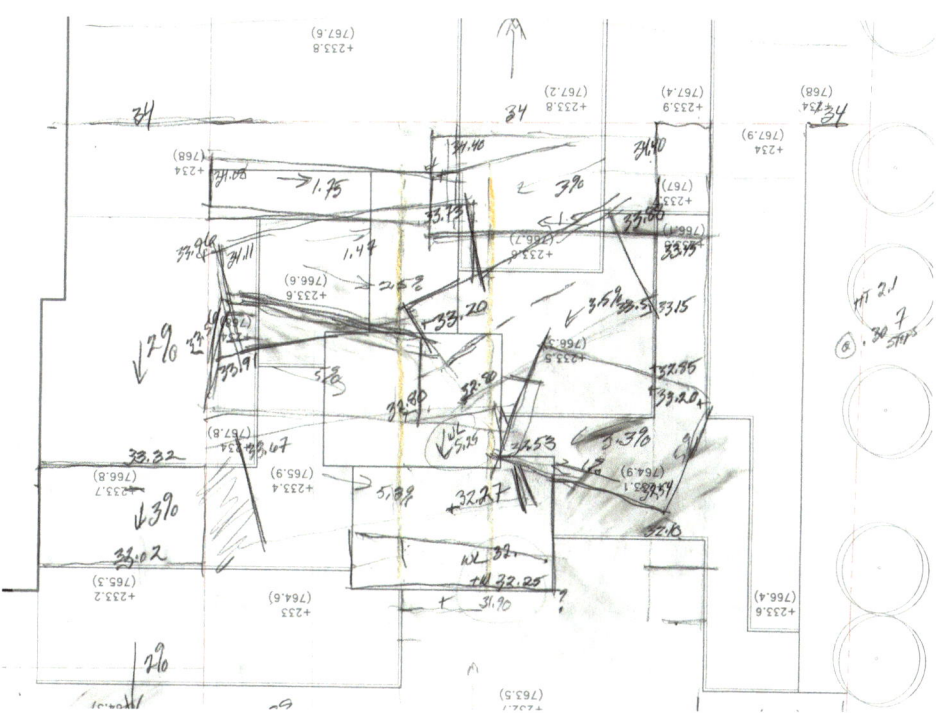

18.4
Detailed grading plan and drawing, Hazelwood Green, by GGN.

18.5
Clay model in process, Hazelwood Green, by GGN.

18.6
3D print of water feature with pencil annotations, by GGN.

Notes

1 The connection between Kathryn Gustafson's sculptural approach to landform and the clay modelling process is frequently cited in writing on her work including Aaron Betsky's essay "The Long and Winding Path: Kathryn Gustafson Re-Shapes Landscape Architecture." In: J. Amidon (Ed.). *Moving Horizons: The Landscape Architecture of Kathryn Gustafson and Partners*. Basel, Switzerland: Birkhäuser, 2005, p. 7.
2 Scolari, M., Ackerman, J.S., and Dussek, J.L. *Oblique Drawing: A History of Anti-perspective*. Cambridge, MA: MIT Press, 2015.
3 In particular section ii of Panofsky, E. *Perspective as Symbolic Form*. New York, NY: Zone Books, 2012. Bottom of Form.
4 Conversation with the author.
5 Snyder, J., Silver, L., and Luttikhuizen, H. *Northern Renaissance Art: Painting, Sculpture, the Graphic Arts from 1350 to 1575*. Upper Saddle River, NJ: Prentice Hall, 2005, p. 386.
6 Here the John Beardsley's description of the picturesque as more cinematic that pastoral applies in Beardsley, J. *Earthworks and Beyond: Contemporary Art in Landscape*. New York, NY: Abbeville Press, 2006. Jane Amidon directly connects the picturesque to Kathryn Gustafson's work in her introduction "The Landscape Architecture of Kathryn Gustafson and Partners." In: J. Amidon (Ed.). *Moving Horizons: The Landscape Architecture of Kathryn Gustafson and Partners*. Basel, Switzerland: Birkhäuser, 2005, p. 89.

References

Amidon, J. *Moving horizons: The Landscape Architecture of Kathryn Gustafson and Partners*. Basel, Switzerland: Birkhäuser, 2005.
Beardsley, J. *Earthworks and Beyond: Contemporary Art in L*. New York, NY: Abbeville Press, 2006.
Panofsky, E. *Perspective as Symbolic Form*. New York, NY: Zone Books, 2012. Bottom of Form.
Scolari, M., Ackerman, J.S., and Dussek, J.L. *Oblique Drawing: A History of Anti-perspective*. Cambridge, MA: MIT Press, 2015.
Snyder, J., Silver, L., and Luttikhuizen, H. *Northern Renaissance Art: Painting, Sculpture, the Graphic Arts from 1350 to 1575*. Upper Saddle River, NJ: Prentice Hall, 2005.

19 The Eidetic Drawings of James Corner

Tina George and Nadia Amoroso

James Corner's Field Operations is an award-winning landscape architecture and urban design firm with over three decades of international recognition. With notable works such as the High Line in New York City, Tongva Park in Los Angeles, the Underline in Miami, and Domino Park in Brooklyn, James Corner has paved the way for a new generation of creative practice. As Professor and Chair at the University of Pennsylvania School of Design from the early 1990s to 2012, his academic background has allowed him to push the profession's norms and the visualization of landscape ideas. In addition, Corner's unique visual communication style has inspired us to 'see' landscape visions as 'eidetic' imagery, possessing conceptual content as well as atmosphere and appearance.

In 1990, Corner started teaching at the University of Pennsylvania and became Professor and Chair of the graduate Landscape Architecture program in 2000. Corner brought a dynamic and innovative approach to the program by championing design creativity, ecology, innovation, and experimentation. He was interested in mapping, collage, and experimental imagery as a form of expression to foster landscape ideas and advance design research. In doing so, Corner's work has been viewed as evocative, innovative, unique, and visionary. It is an adventure to interpret his drawings and ideas. One needs to 'peel back' layers of information that seem to push and reverberate through the spaces of the manifested piece, begging to be discovered. Such is the power of his suggestive imagery and his visual expertise, which also plays out in his actual design projects.

In the early days, Corner is best identified by his formulation of 'map-drawings,' a foray into datascapes, endeavouring to capture the quantitative and qualitative characteristics of the space. He does so by depicting analytical information on a flat-bed plane while embracing high aesthetics in his representation. This eclectic mix of information and aesthetics seems to recrudesce in one's mind and recalls vividly, never to be easily forgotten.

Take, for example (Figure 19.1), fittingly named 'Windmill Topography.' Corner avidly portrays the best of composite montage, incorporating multiple layers and types of information by re-introducing the McHargian method of overlaying information (using maps, pictographs, tracings, data, etc.).[1] MacLean's aerial photo acts as a base to display key pieces of data such as wind directions, wind pressures, ridgelines, contours, and text depicting the place, LA (Los Angeles). The frame is open and airy, and the viewer's vision flows in and out of the white spaces of the drawing, much like the windmills themselves, capturing and absorbing the flowing wind currents. The picture of the actual site is situated to the bottom right corner allowing a peek into the textural ecology of the land, replete with the mountainous terrain, which duplicates itself in the form of a horizontal black strip taken from a section cut of this terrain. The viewer immediately recognizes the anchoring of the windmills onto the mountainous terrain and the importance of the massiveness of the landform upon which they are situated. The windmills themselves make multiple appearances throughout the drawing; they punctuate the piece at different scales and are even scattered in profile. This multiplication of windmills against a carpet of terrain seems to direct the viewer's attention to the understanding of the site—little windmills in an unruly, colossal land mass, nevertheless casting a significant impact.

DOI: 10.4324/9781003183402-19

19.1
Windmill topography. Exhibiting data associated with wind pressure and air temperature. By Corner et al. Adapted from *Taking Measures across the American Landscape*, Yale University Press, by Corner, J., and MacLean, A., 1996, p. 83.

The use of intense saturation is limited to the contour map and the section cut, riddled with thin red lines indicating wind directions. These two saturated elements are offset with the rest of the canvas left blank, encouraging white space to balance the piece. The data and imagery are presented in an engaging manner, calling the viewer to decipher and search for more information within the drawing. Another exciting element is the arc-shaped cut-out from the contour map that lies on the bottom left just below the section cut. This technique seems to be a throwback to how the images were originally captured for the Corner–MacLean project: light airplane photography captured by a Cessna 182 flown by MacLean. The propeller of the Cessna 182 in motion takes on a shape like this small arc. The theme of airplane elements is found to be a recurring one in some of Corner's works.

Ultimately, Corner's many map-drawings collated in the seminal *Taking Measures Across the American Landscape*. This book represents a process of collection and juxtaposition, culminating in collages that encourage interpretation and dialogue, avoiding mere 'direct' imagery. Essentially, the limiting of human imagination with ready-made realistic images defeats the purpose of the need for the mind to engage and be part of the process of seeing the world with new eyes, and opening up new possibilities for creativity. The only reference to explaining the drawing is the supporting caption below the image; the rest is an open question, inviting interpretation.

Karen M'Closkey shares Corner's views on collage as a highly effective medium to synthesize a design representation that uses systemic montage, the flat-bed being the backdrop for putting down ideas. These eidetic images evoke dialogue and are rich in data that is 'seen but unseen.' In doing so, the resultant design presents a new way of thinking to the viewer and forgoes imitation.[2]

The Eidetic Drawings of James Corner

As both an academic and practitioner, Corner continues to participate in international design competitions as a means of experimentation and international development. In the early 2000s, James Corner, teamed with architect Stan Allen, became one of the finalists for the seminal Downsview Park Competition in Toronto. Corner's imaginative collages captured the idea of 'emergent ecologies' within a large urban park, a dynamic, evolving concept as opposed to anything final and fixed. His perspectives alluded to both the history and future outlook of the site.

In Figure 19.2, the collage 'Cultural Campus Event Fields' depicts a group of students sitting and picnicking atop an earth-hill, taking in a spectacular view of the amphitheatre, the hangar, and the playing fields. It is relevant to mention that Downsview Park is a closed military airbase, and it is fitting to see the detail that Corner places into the collaged image—a black and white fighter plane that 'hangs off' the top right-hand frame, a black and white hangar seen in the distance—all reminiscent of things that 'have been'; collecting these cultural memories into the fabric of the design seems to be something that Corner is dedicated to. This representation is a key example of the urban renewal/reclamation project that Field Operations values highly in their project profile. Similar to 'Windmill Topography,' the scale of the project is large, and the imagery is captured from a far distance, allowing the artist to display the sort of programming that is key to the project itself—in this case, a circuit of pathways and spaces for activities. Yet another interesting feature that throws back connections to Corner's previously mentioned collage is an alignment of wind turbines that bank the image's background. Immersed in trying to capture the right ecological solutions and wind being an important factor in site selection for airports, this element seems to be at ease in the picture. The sense of wind blowing through the space also creates a light, airy, and leisurely paced energy within the perspectival view. The images of the students and the plane project out of the frame ever-so-slightly, as if they are in motion and want to break free of the landscape. A feeling of needing to explore beyond borders is evident, and just like the landscape that Field Operations creates with its emergent successional qualities, these elements also seem to 'emerge' from the frame. The image itself lacks the quality of a fully developed design owing to its grainy texture, allowing it to ingrain a softer memory in the viewer's mind. It also reminds that Corner's imagery is dynamic and open to changes, and as

19.2
Cultural campus event fields. An activity track flows through the site as a circuit, and an earth-hill provides a spectacular view of the activity fields and the amphitheatre. By James Corner, Stan Allen, and design team for the Downsview Park competition, 2000.

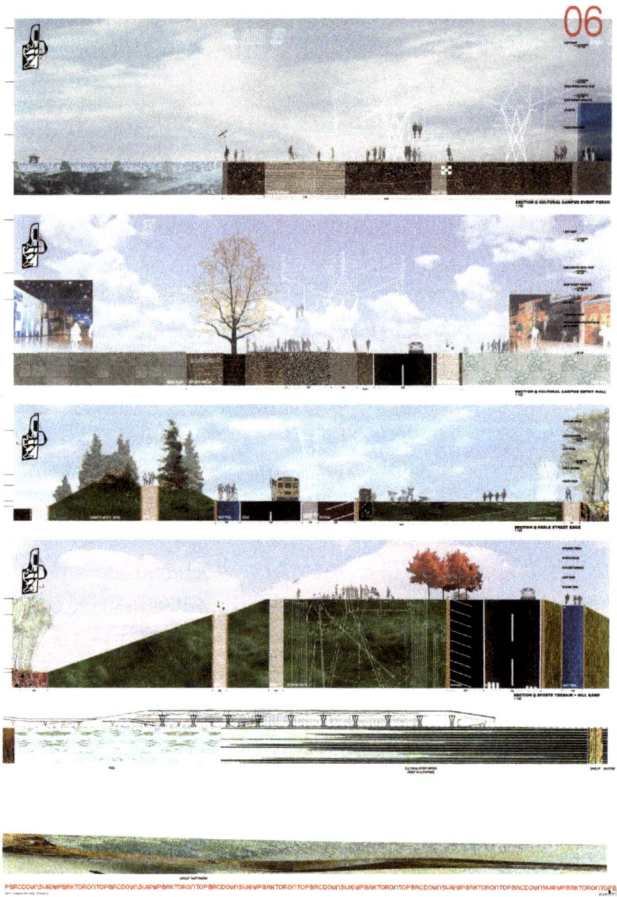

19.3
Downsview section cuts. Focusing on the textural and aesthetic qualities of the ground plane and the interaction of the viewer with the space in relation to nearest objects. By James Corner, Stan Allen, and design team for the Downsview Park competition, 2000.

with other of Corner's drawings, there is no text or annotation in the image, except for the supporting caption below the image.

On the other hand, while Corner's perspective drawings are intriguing, his combination of section drawings with the partial plan below the cut-line allows the drawings to tackle two narratives at once (Figure 19.3). This dual drawing type becomes a clever approach indirectly connecting the vertical dimension with the ground plane. Experimenting in this manner is a new approach for Corner, which is brilliantly executed. It highlights the textural values of the space and the relationship of the viewer/explorer with nearby elements. It is worth mentioning here that fields of texture seem to be a big part of Field Operations and have been raised as a valuable component in Corner's designs; they allow the design to be holistic and temporal while creating the greatest impact.[3] However, Figure 19.3 is not just another typical section cut. Corner puts his spin on the imagery by incorporating data within the image (especially in section cut 4, within the grass field) to bring a sense of justification to the design. It suggests to the viewer that the design has been prepared after research and is legitimate. In doing so, Corner's experimentation in the visualization of the section-plan offers a sense of reasoning while increasing speculation, wonder, and curiosity for the space. The visual communication of his work becomes a critical aspect of his design process.

The Downsview Park project was also instrumental in the development of Corner's supremely distinctive imagery of seasonal and successional timelines (Figure 19.4a). These types of ecological and seasonal time-lines diagrams are highly appreciated by landscape architects, and these 'types' of diagrammatic charts are used effectively by many landscape architects. This kind of imagery is an example of the acute mannerism of Corner's thought process. He elevates landscape from a rudimentary

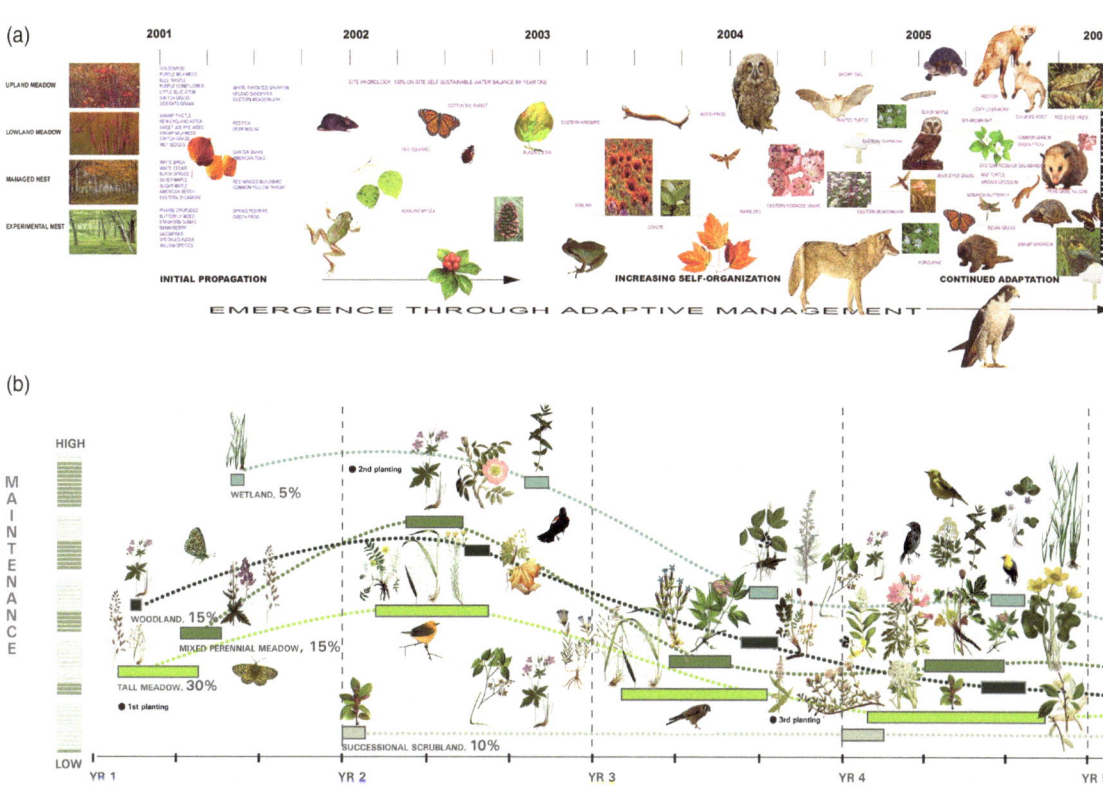

19.4
(a) Ecology timeline chart: emergence through adaptive management. Depicting ecological succession through a stepped timeline, by James Corner Field Operations with Nina-Marie Lister, Downsview Park, Toronto, Canada, 1999. (b) High Line ecology timeline diagram. Depicting biodiversity and level maintenance over time at the High Line, by James Corner Field Operations, High Line Competition, 2006.

aesthetic to a constituency of various factors including terra, flora, fauna, memories, and nodes that are desirous of social interaction. This mapping technique also collects the level of flux that the land beholds; it is not stagnant but moving and evolving along both axes. Reading the drawing culminates in a deep awareness of time and persuades the viewer to think beyond the 'now.' Corner hints at the ephemeral qualities of site and allows the unseen to be envisioned through the simple act of inscribing images on a timeline ecology chart. This clever style of image imposition allows the viewer to comprehend the complexity of the site which cannot be captured by a simple perspectival image, thus quenching the standard desire to see a photo-realistic image. It is interesting to note that, in keeping with some of his other works, there is reference to the river birch tree, data is noticed imprinted within the frame, and there is a sense of emergent movement in the tone of the image. This form of expression is simple but powerful and has been replicated numerous times because of its effectiveness. In generating the timeline imagery, Corner has truly set the standard in design expression of this style. In fact, the application of the timeline graphic finds both aesthetic as well as academic application. In a recent article, Corner has alluded to his quest of capturing the mental map and his aspiration to design with the mind in mind.[4] The timeline graphic of Downsview Park secures this intention thoroughly and successfully.

One of Corner's most acclaimed projects that cemented his branding and garnered international recognition would be the High Line project in New York, which commenced in 2006. Yet another large-scale project, Corner worked with architecture firm Diller Scofidio + Renfro and planting designer Piet Oudolf to deliver this stunning landscape. It was another urban renewal/reclamation project like that of Downsview Park. Rediscovering meaning within the site using found objects galvanized his design process just as much as the innovation of design to deliver a radically new way of experiencing the site. Again, Corner utilizes a timeline ecology diagram to visualize the biodiversity

19.5
View of the grasslands. The High Line, New York City competition drawing, by James Corner Field Operations, 2006.

of planting and wildlife on the High Line over time (Figure 19.4b). This visualization is clear and powerful, allowing the viewer to understand the ecological diversification and level of maintenance over time.

Figure 19.5, 'View of the Grasslands,' is a collage that Corner created early on in the process to help shape the design concept. Here, one can see the simple juxtaposition of crucial elements that play into the experience of the site, blending existing context with new interventions, both hard and soft. Corner keeps it simple and easy to read. The result is an incursion into the site experience rather than just a highly curated piece of art. Corner places paramount significance on the textures and colours. The found objects such as the rail tracks are weaved into the design, and the planar view is lifted into the horizon. The design experiments with the qualities of hardscape and softscape blending into each other, and the sightline is focused on the near and now. Attention is placed on the human scale of being in the space. The High Line design evokes simplicity in its tone and form. It takes advantage of the views over the city using street-viewing 'spurs,' 'balconies,' and 'frames' jutting out dramatically from the linear pathway. Once again, Corner utilizes the aerial vantage point as a design element to perfection. In his drawing, the buildings flanking the path are not photoshopped and are left untouched so that the viewer has a 'real' sense of the experience of the design. The frame of the drawing is only partially altered with some features that are the crucial elements of design. At the High Line, Corner is also starting to create modular paving, which comes together in pieces. This technique ends up being a clever way of adapting the design and making it feel less strict and more leisurely, light, and flirty. In the background is a grove of undoubtedly thriving birch trees. They lend an air of lightness to the design, promising dappled shade without the intrusive boldness accompanying larger shade trees. It is worth mentioning that birch trees appear in much of Corner's imagery. Birches are aesthetically pleasing trees, with paper white trunks and a light, airy canopy. They work well in urban conditions and are ecologically appropriate to urban environments. Even in the most unobtrusive design like that of Figure 19.3, in section cuts 2, 3 and 4, there are hints of the light silhouettes of birch trees that draw one in.

The above examples of Corner's works ride on the experience of public engagement. Design is kept light-hearted, mutable, and breezy in texture. Figure 19.6, an eidetic photomontage for a project proposal in Stockholm (1999), presents a highly speculative and imaginative sense of the place. This image is very intimate, palpable, and rich, engaging one's full senses. The engagement is with a singular person, enveloping the audience in a landscape that is raw and unapologetically real. There is a palpable sense of 'tryst' with every single element in the collage, all highly saturated to reveal the high intensity which the interaction is revealed to be. The birch trees are stark white and display ardent peeling, as if in conjunction with the nude woman. This indicates an

19.6
Eidetic photomontage, Stockholm. This image employs generalized depiction of place, rough extractions, and an ambiguous frame. By James Corner, 1999.

undeniable connection between humans and the land, as if one follows the other. Images of the ground texture are luscious, exaggerated, and unpredictably scaled. The image captures a rendezvous between the human and the landscape. Such is the power of colour and scale that Corner brings to his imagery; reading this drawing is an evocative exercise, pulling at the memories tied to land and water. One can almost feel the crunch of leaves underfoot and the thick wind from the bay rushing across the skin. This eidetic imaging burns into one's mind and Corner does this best with his disregard of a rectangular frame, pieces of images projecting into the white space, unabashed and untethered.

Figure 19.7 illustrates one of Corner's more recent collages. The Camden High Line in London competition drawing (Figure 19.7) further strengthens Corner's creative drawing style. Similar to the New York High Line, the Camden High Line project (2021) is both large and intimate in scale and carries a very strong aura of place and context. The same sort of human–nature interactive design supporting urban renewal/reclamation is on view. However, the tone and saturation are vividly different from the New York High Line project. In delving deeper, there is a higher gradient and variety of interaction within the Camden Project, which implies a more diverse set of programming parameters. In this design script, the viewer's eyes are also led to look up, not just at the horizon, but also at the verticality of the space—a found element of railway portals is used to raise the sightline upward instead of just linearly (as was with New York's High Line). Some of the elements from Corner's earlier imagery are well juxtaposed herein; one can find the birch trees with their peeling bark, found objects, and the well-thought textural effects of the ground plane within this drawing. Balconies similar to the New York High Line are continued in this design as well. It starts to differ in the levels of saturation and the textural qualities of the accompanying ground vegetation. In Camden, there is an abundance of fruit and flowers, more relatable to the warmer English climate and factor in the fresh and farm-food market programming. In addition to these elements, the higher saturation emulates the found graffiti on the walls. With the Camden Project, Corner mimics his well-placed attempt from the New York High Line for modular pathways, this time using a herringbone wood planking system. The framework is jagged and ambiguous in the proper fashion of Corner's earlier works, suggesting that design exploration has not ended but is ever open to change.

In Corner's illustrations, definite design elements cater to the audience that it is supposed to serve. There are tone and saturation differences that increase with the magnitude of intimacy with

19.7
Towards growing Camden's New Green High Line competition drawing, by James Corner: Field Operations, 2021.

the landscape and decrease with the rise in human interaction. However, the similarities outweigh the differences, making Corner a figure to be reckoned with. He displays a keen sense of artistic branding with the repeated elements in his imagery; he stays true to his mission of eidetic imaging to serve the body and mind, and his images function dynamically as suggestive and unfinished collages, open to creative interpretation.

Something that is not as well-known about Corner is his aptitude for trepidatious sketches. Much like his flat-bed imagery and collages, they portray a tone of urgency and motion. And yet, even his field sketches shown in Figure 19.8 have a level of finesse to them, as can be observed in the detail of the birch on the left side of the drawing. Corner seems to choose spaces of intense artwork in his drawing versus areas with milder brushstrokes and balances out the drawing delightfully. The concept of prospect and refuge discussed in his article, "Mind Landscapes: Navigation, Habitat and Imagination"[5] is sought after and captured in this drawing by enclosing his subject in a thicket of underbrush, looking onto a wide open vista ahead of them.

By borrowing from past landscape architecture concepts such as the McHargian theories and Appleton's prospect and refuge theory, Corner has truly enlightened us with art that has meaning and application while still being aesthetically pleasing. In studying his artistic pieces, there is a sense of understanding that Corner indulges his appetite for making change across large complex landscapes and elevating human-nature interaction with the right mix of elements. Considering that these chosen landscapes are ephemeral and fluid in their ecological processes, inevitably changing and growing with time, it begs the question: *Is this why collages work well for these living and open-ended sites?* Corner seems to have made that connection by creating works of art that are just as unpredictable and open-ended as their actual design intent. Both the imagery and the projects themselves have fluid and dynamic properties; they are at once spatial and tactile while pointing to imaginative sets of possibility and interpretation.

It also seems that the palpable unpredictability that navigates the edges of a large living landscape, which continually changes over time, might better be served by the art of mapping and collage to better approximate its soul or *genius loci*. Corner shows us how to do this (and more) just right, served with a dash of provocation!

19.8
Field sketch. Graphite on paper, by James Corner, early 2000s.

Notes

1 Corner, J. "Eidetic Operations and New Landscapes." In: J. Corner (Ed.). *Recovering Landscape: Essays in Contemporary Landscape Architecture*. New York, NY: Princeton Architectural Press, 1999, p. 166.
2 M'Closkey, K. "Structuring Relations: From Montage to Model in Composite Imaging." In: C. Waldheim and A. Hansen (Eds.). *Composite Landscapes: Photomontage and Landscape Architecture*. Ostfildern, Germany: Hatje Cantz Verlag, 2014, pp. 116–131.
3 Phaidon. "James Corner's High Line Vision, 2006." June 22, 2021. https://www.phaidon.com/agenda/architecture/articles/2015/september/03/james-corners-high-line-vision/.
4 Corner, J. "Mind Landscapes: Navigation, Habitat and Imagination." *Architectural Design*, 2020, 90(6): 60–65.
5 Ibid.

References

Corner, J. "Eidetic Operations and New Landscapes." In: J. Corner (Ed.). *Recovering Landscape: Essays in Contemporary Landscape Architecture*. New York, NY: Princeton Architectural Press, 1999, pp. 152–169.
M'Closkey, K. "Structuring Relations: From Montage to Model in Composite Imaging." In: C. Waldheim and A. Hansen (Eds.). *Composite Landscapes: Photomontage and Landscape Architecture*. Ostfildern, Germany: Hatje Cantz Verlag, 2014, pp. 116–131.
Phaidon. James Corner's High Line vision, 2006. June 22, 2021. https://www.phaidon.com/agenda/architecture/articles/2015/september/03/james-corners-high-line-vision/.

20 Non-sites and Simulacra

Ken Smith

To be quite honest, my graphic representation style results from not being very good at freehand drawing. I was never one to go around with a pad and pencil sketching. I was more aligned with television and photography. I got my first camera when I was in junior high school, a 'Kodak Brownie' box camera, which was a very basic camera that used 120 mm film, and I took my first photography classes in 4-H club. In high school, I inherited my father's Argus c3, a 35 mm camera, and in college I bought a Nikon FT2 single lens reflex camera. I am a largely self-taught photographer and I also learned darkroom processing and printing while in college. Therefore, using photographic images comes naturally to me in producing graphic representations in my work.

As an undergraduate student, while studying landscape architecture at Iowa State University, I was also taking elective art and sculpture classes. I spent considerable time roaming the open art stacks in the university library, mostly interested in Modernism, Constructivism, Dada, and Surrealism, and reading the dense art-practice-theory articles on Conceptualism, Minimalism, and, in particular, the Earthwork artists in the bound volumes of Art Forum magazine. I continued these interests in graduate school at Harvard's Graduate School of Design (GSD), developing ways to combine my interests in contemporary art with the practice of my professional discipline.

When I opened my own office in the early 1990s, most graphic production was still pre-digital. My graphic representation practice at that time was mixed-media, combining traditional pencil and ink drafting with paste-up illustration materials such as Letraset, Chartpak graphic tape, and 'Zip-A-Tone' film, in combination with photo prints, photocopies, colour laser prints, and magazine and other clip art materials. It was a very fluid way of working that allowed for experimentation and often resulted in interesting juxtapositions of material and content. You might say my photo-based technique was more X-Acto knife and adhesive than pencil and paper. What appealed to me about this method of production was that it had discernible edges and, while made of mediated material, it still showed the handwork in its making. These early montages were precise in image but still abstract; they were simultaneously real and synthetic, grounded and conceptual. My photo montages are in the tradition of ideograms—they are not intended to be finished, realistic, or accurate three-dimensional representations. They are pictorial but not picture-perfect. They are a shorthand representation meant to communicate an idea and a sensibility. They are a concept statement in visual form.

It is useful to place this pre-digital graphic production and montage representation within the post-modern art context of the time. Living in New York City, I had ready access to the currents in the art scene and my graphic montage work reflects that spirit and influence. For me, this type of graphic production and landscape representation was entirely in line with the work of Robert Smithson and his notion of Non-Sites, which he described as a "logical picture that is abstract, yet it represents an actual site."[1] He wrote that a "logical picture differs from a natural or realistic picture in that it rarely looks like the thing it stands for. It is a two-dimensional analogy or metaphor."[2] I was reading Smithson and seeing his work at the same time I was reading Jean Baudrillard's notions of simulation and simulacra,[3] and seeing the work of the Neo-Geo, Pictures Generation and image appropriation artists in the SoHo and Chelsea galleries. The works of John Baldessari, Peter Halley, Barbara Kruger, Sherrie Levine, and Richard Prince were of particular interest. The influence of post-structuralism was affecting all creative fields. Collectively, all of this had an immediacy and impact on the images I was creating as part of my professional practice. The art and creative worlds in the late 1980s and 1990s were very free-wheeling with a lot of experimentation and norm-busting. It was a time of challenging precedents, developing new modes of artistic and design expression, and being simultaneously subversive, ironic, and playful. The practice of landscape architecture was also changing. This was a time when there was a blurring of professional boundaries with a lot of crossover collaborations, a mixing of high and low culture, mixing of various media, and a recycling of past styles. It was also a period that was open to art-oriented practice, and there was a re-emerging environmental and social movement underway in reclaiming public space in urban areas. Being able to communicate graphically in a contemporary way was a definite advantage in advancing art and design ideas.

In the 1990s and early 2000s, emerging digital technologies were beginning to profoundly change the professional practice. To be successful, firms needed to work digitally and collaboratively. Photoshop came very naturally to me since it used many of the same techniques as photography

20.1
Emerald Rules, 2005, and remade as "Portable Garden Stourhead," 2020, 11.5 × 14. Early in my practice, I formulated a set of 'emerald rules,' both tongue-in-cheek and serious, to define my practice. One of them was portability, and I made a montage of an eighteenth-century gardener pushing a cart containing a garden. In 2020, I remade the image inserting a view of the eighteenth-century Stourhead garden in the labourer's cart. While I could have more easily remade the image in Photoshop, I chose to recreate it as a cut-and-paste montage on acid-free paper, along with six other new hand-made montages exploring similar themes of garden portability, the idea of non-site and landscape as container of cultural expression. By Ken Smith Workshop.

20.2
Brooklyn Bridge Garden Mount, 1990, 21 × 25. This is an early example of my montage graphic representation approach done for a competition to improve the connectivity in and around the Brooklyn Bridge approach ramps. This proposal appropriates the historical form and concept of a 'garden mount' and transforms it at the scale of an American earthwork. The background is a photocopy enlargement, and the garden mount graphic is an original pencil-on-vellum drawing reproduced as a high-contrast black and white silver gelatin photo contact print. The photo is rendered with Pantone colour translucent film to give it a flat, even colour. By Ken Smith Workshop.

darkroom manipulation, such as burning and dodging, masking, perspective plane shift, brightness and contrast change, and multiple exposures. Digital cut and paste tools made montage easy. My grounding in photo-based montage, appropriation, and mechanical reproduction made the transition to digital work in Photoshop and other digital graphic platforms a very natural change, and the digital and synthetic representation of landscape imaging is what typifies the work in my office now. Almost all aspects of contemporary design production today are digital and often remote, but I find that the skills and practices of image-based montage are still effective tools in graphic communication. Figures 20.1–20.7 are a collection of my 'drawings' representing landscape ideas, using montage techniques. The image captions describe the visuals and techniques used.

Non-sites and Simulacra

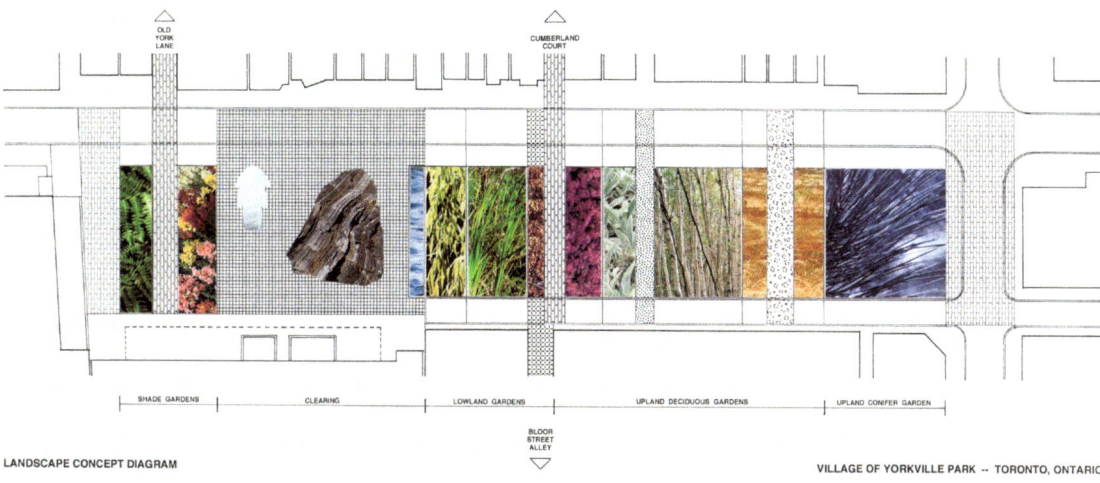

20.3
Village of Yorkville Park Concept Montage, 1990, remade 1996, 18 × 24. This graphic started as a hand-drawn, pencil-on-vellum plan with Letra-tone graphic pattern film. It was converted to a high-contrast black-and-white photo print and montaged with colour clippings from a landscape picture book. The approach is simultaneously realistic and abstract in communicating the concept of a collection box garden set within an urban context. By Ken Smith Workshop.

20.4
A Long Look at Fifth Avenue, 1997, 27 × 39. This montage was done entirely with standard-size Kodachrome colour prints mounted on acid-free art paper. The background was montaged together from multiple site photos, and the chandeliers were photos taken in the New York Public Library reading room. This image was created for an invited re-visioning of the avenue sponsored by the Municipal Art Society. I had concluded that the buildings, sidewalks, and streets along the avenue were fully occupied, and the best remaining landscape opportunity was to address the ceiling of this public space. By Ken Smith Workshop.

Hair Garden Type One:
BOUFFANT TOPIARY, puff out that hair, let it grow and prune it into fantastic ornamental forms.

20.5
Hair Gardens Type One, Bouffant Topiary, 1998, 11 × 17. This is from a series of eight montages exploring the parallels between landscape and fashion design. Using pages from fashion magazines and clip art from garden magazines, this series explored personal expression and fashion dictates of taste with common themes of social display, ideology, desire, sex, power, pretense, ethnicity, and health. Appropriating both the imagery and presentation style of fashion magazines, the hair garden proposals include Bouffant Topiary, Ear Hair Rockery, and Chest Hair Maze. By Ken Smith Workshop.

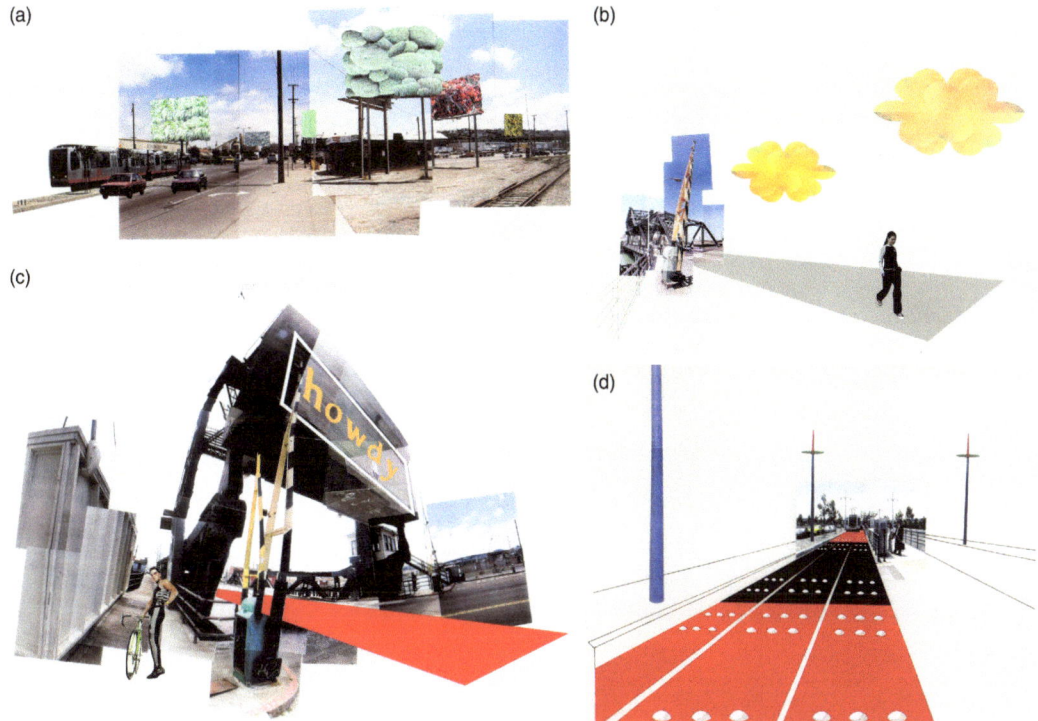

20.6
A–D Third Street Light Rail, San Francisco, 1998, 36 × 48. There were 11 montages in this series of concept images for a public art commission to be installed along a six-mile transit corridor south of downtown San Francisco in the Mission Bay, Dogpatch, Central Waterfront, and Visitacion Valley neighbourhoods. The montages combine ink drawing on mylar with photographs and other clip art, graphic paper, and film. The four montages shown here, Topiary Billboards, Thought Bubble Lights, Mission Creek Channel Gateway, and Distinctive Trackway Paving are typical of the set and were presented to the San Francisco Art Commission and MUNI municipal railway. Ultimately, my part of the public artwork was the design of six miles of distinctive trackway paving. By Ken Smith Workshop.

20.7
Hanging Gardens of Queens, 3D lenticular print, 2011, not to scale. This is a Photoshop before-and-after set of images printed with lenticular technology such that the images shift from before and after from different viewing angles. The images contrast a photograph of found billboards proclaiming 'Drugs', 'Gangs,' and 'Dropouts' replaced with living vertical topiary billboard structures. This one hung in the front hallway of my loft for many years and appeared different depending on whether one was entering or leaving. By Ken Smith Workshop.

Notes

1 Smithson, R. "A Provisional Theory of Nonsites." In: J. Flam (Ed.). *Robert Smithson: The Collected Writings*. Berkeley: University of California Press, 1996, p. 364.
2 Ibid.
3 Baudrillard, J. *Simulacra and Simulation*. trans. S.F. Glaser. Ann Arbor: University of Michigan Press, 1994.

References

Baudrillard, J. *Simulacra and Simulation*. Trans. S.F. Glaser. Ann Arbor: University of Michigan Press, 1994.
Smithson, R. "A Provisional Theory of Nonsites." In: J. Flam (Ed.). *Robert Smithson: The Collected Writings*. Berkeley, CA: University of California Press, 1996, p. 364.

21 The Spirit of Drawing

Chip Sullivan

Throughout my career as a landscape architect, my belief in climate-conscious design has been steadfast. For more than three decades, I have been committed to creating landscapes that conserve energy and likewise function as works of art and places of spiritual renewal. The garden has been my teacher; drawing, my inspiration; and art, my form of expression.

Trial by Fire: Weekend Watercolour Excursions

After graduating from the University of Florida, I went to work for Sasaki Associates in Coral Gables, Florida. I put my art background to use in the office and produced many presentation drawings and renderings. Being perceived as the resident artist, several of my co-workers asked me if I would teach them how to watercolour. I had no idea how to watercolour, but out of fear of embarrassment, I said, "yes!" I rushed around to bookstores to clandestinely study books on watercolour painting, as I had very little money to purchase books. My mother had bought me a Winsor & Newton Cotman travel watercolour kit and a small pocket brush to encourage me to paint. She even sent along a small plastic container for water. I bought a Sennelier 4 in × 6 in *album a dessin* sketchbook and prepared to give my first lesson.

My colleagues and I decided to alternate visits to Villa Vizcaya and Fairchild Tropical Gardens on Saturday mornings. The Villa Vizcaya was built in Miami between 1914 and 1922 as a subtropical interpretation of an Italian Renaissance garden. Fairchild Tropical Gardens was a unique botanical garden in Miami based on a formal masterplan. Both sites had abundant subjects to paint. Our first excursion was to Villa Vizcaya, and thus, I began my trial by fire. Somehow, I was able to 'fake' a knowledge of watercolour painting by developing techniques intuitively on site. Thankfully, no one guessed that I did not know what I was doing! I used an Eagle 314 Draughting pencil to lay out the drawings and added quick, light watercolour washes.

Refer to Figure 21.1a, A Cool Grotto. Taking shelter from the hot Miami sun, I found respite in an underground grotto. Water dripped down the sides and splashed onto the floor; the cool interior contrasted with the hot and humid garden outside. Refer to Figure 21.1b, Shelter from the Wind. On a cold, windy day in December, I wandered into a sun-filled pocket and painted a sectional study. I later learned that this garden element was called a '*giardino segreto*,' a perfect warm spot for blustery days. I noticed during these visits that I would navigate to areas that were pleasant to sit in and paint, drawn to spaces with comfortable microclimates. This observation was my epiphany! I was driven to research and develop a vocabulary of historical garden elements that moderated the microclimate. My goal was to write a book on passive microclimatic design.

A Creative Paradise: The American Academy in Rome

As an undergraduate, I had seen a poster promoting study at the American Academy in Rome. The *Prix de Rome* fellowship seemed to be the perfect venue to provide the time and space to complete

DOI: 10.4324/9781003183402-21

21.1
(a) A Cool Grotto. By Chip Sullivan. (b) Shelter from the Wind. By Chip Sullivan.

my research. After the heartbreak of two unsuccessful applications, I was thrilled to be awarded the Rome Prize on my third try. The fellowship allowed me the opportunity to examine firsthand the passive design elements in the great gardens of history. Armed with a fancy new watercolour kit from the Zecchi Colori Belle Arti store in Florence, I began my studies of site context and climatic design in a 5 in × 8½ in Vang sketchbook. Yet, after drawing and analyzing hundreds of design features of historical Italian Renaissance, Moorish, and Islamic gardens, I had no idea how to organize my extensive collection of material.

Refer to Figure 21.2a, Villa Rotunda, Vicenza, Italy, 1566–1569. An afternoon visit to Palladio's Villa Rotunda in Vicenza inspired me to structure my volumes of research drawings into four sections. This watercolour sketch of the view from the west shows the relationship between the villa's loggias and the woods (*boschi*) along its southern edge. Refer to Figure 21.2b, The Rotonda's Orientation. This site plan illustrates the villa's solar orientation, its relationship to the cardinal directions, and individual views. The cross-section depicts the microclimatic effect of the *giardino segreto* (the secret garden)

21.2
(a) Villa Rotunda, Vicenza, Italy, 1566–1569. By Chip Sullivan. (b) The Rotonda's Orientation. By Chip Sullivan.

within the bosco. The diagrammatical section shows the villa's water collection and storage system. The four loggias each face a different direction and frame a view of a distinct landscape: a bosco to the south, agricultural fields to the east, the Bacchiglione River to the north, and a sweeping landscape panorama to the west with Vicenza as its backdrop. Interestingly, the villa was purposely rotated 45° from north, so each loggia would be filled with sunlight at a different time of the day. The Villa Rotonda orientation suggested to me an idea for organizing my research according to the four cardinal directions and the four Platonic solids—Earth, Fire, Air, and Water. As my fellowship at the American Academy in Rome was coming to an end, I finally developed a very rough outline for a book proposal on climatic design.

A Design Vocabulary for Sustainable Landscapes

When I returned from Rome and began teaching in the landscape architecture department in the College of Environmental Design at the University of California, Berkeley, I was able to refine my drawings and draft the manuscript for *Garden and Climate*. My book proposal was accepted and published by McGraw-Hill. Pen and ink have always been my favourite drawing medium. I love the feel of the highly responsive nibs that produce expressive lines. The dip pen makes a pleasant sound as it grazes the paper. Inspired by the delicately detailed pen and ink drawings in Hubbard and Kimball's vintage book, *An Introduction to the Study of Landscape Architecture*, I decided to render the illustrations for *Garden and Climate* in pen and ink.

The concept sketches were developed on Borden & Riley Sun-Glo Thumbnail Sketch Paper. I produced many overlays. When I was finally satisfied with the composition, I laid a sheet of 11 in × 14 in Canson Vidalon vellum over the drawings. This paper has high transparency and an excellent tooth for ink drawing. It also withstands erasing ink lines with an electric eraser for the inevitable mistake. I used an HB pencil to lay out the composition with light pencil lines. As my final and favourite step, I inked the pencil lines with a Speedball crow quill dip pen.

Refer to Figure 21.3a, Vicobello, Siena, Italy, sixteenth century. Along the villa's main terrace is a lemon garden. The solar-powered *lemonaia* borders the garden to the north. I used extensive cross-hatching to create tone and texture. Refer to Figure 21.3b, Villa Torlonia, Frascati, Italy. Two parallel tunnels create cool air pockets and funnel them into the bosco on the top terrace. I drew vertical lines

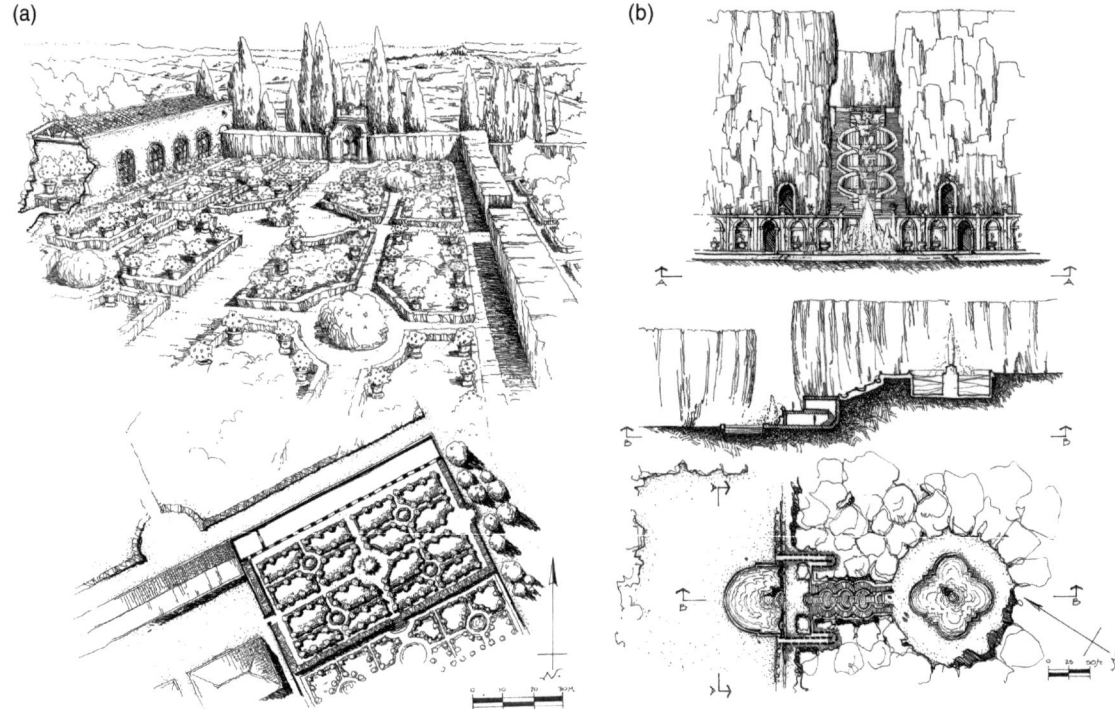

21.3
(a) Vicobello, Siena, Italy, sixteenth century. By Chip Sullivan. (b) *Villa* Torlonia, Frascati, Italy. By Chip Sullivan.

to illustrate the form and depth of the clipped bosco, parallel diagonal lines for the shadows, and dense double cross-hatching for the earth in section.

Beaux-Arts Mash-Up

The final step in preparing the manuscript for *Garden and Climate* was to create prototypical gardens based on my design postulates. I chose to employ classic beaux-arts watercolour techniques. I drew the softer atmospheric lines in pencil to accentuate the watercolour washes. I used mounds of tracing paper to work out the design. Once finalized, I used a light table to transfer the drawings onto a 22 in × 30 in sheet of 140 lb. Arches hot-press watercolour paper. I laid out light construction lines with an HB pencil and then completed the final drawing with an Eagle 314 Draughting pencil. The paper was soaked in water and stretched onto a quarter-inch plywood board by taping along the edges with 1½ in kraft paper tape. When the paper was dry, it would lay flat. For my paint colours, I used a combination of Winsor & Newton pans and tubes. I used Series 7 Winsor & Newton sable hair brushes in sizes 2, 4, and 12. For large washes, I premixed the colour in shallow ceramic bowls. I typically begin painting light broad washes with the size 12 brush and then add details and accents with size 2 and 4 brushes. Once I was satisfied with the painting, I added shadows with a mixture of purple and Payne's grey. When the paper was completely dry, I removed it from the board by cutting along the taped edges. This process results in a flat drawing with no buckling, which is ideal for scanning.

Refer to Figure 21.4a, The Garden of Bacchus. This garden was designed for a site in Napa Valley, about 67 miles north of San Francisco. The climate is scorching in summer and mildly cool in winter. The complex is sited on a south-facing slope with a dense bosco of trees planted on the north side to block the cold northern winds. The garden is open to the south to collect and direct the cool summer breezes into the structure built into the earth to provide insulation. Grapevines and orchards help cool the air through evapotranspiration. Refer to Figure 21.4b, The Garden of the Phoenix. The Garden of the Phoenix was proposed for a site near Taos, New Mexico, a region where the summers are hot and the winters brutally cold. The southern face of the mesa acts as a giant solar

21.4
(a) *The Garden of Bacchus*. By Chip Sullivan. (b) *The Garden of the Phoenix*. By Chip Sullivan.

collector for the building complex. All the interiors have full exposure to the low winter sun for solar heating. Below the complex, courtyards, terraces, and a *giardino segreto* provide a comfortable, year-round microclimate. A series of orchards create a cool shady retreat in the arid landscape.

Garden Reliquaries and the Path to Knowledge

While in Rome, I explored the metaphysical dimension of landscape. I created a series of garden reliquaries, triptychs, and dioramas. As my time in Rome was ending, landscape architect Michael van Valkenburg asked me if I would like to participate in an ideas competition titled *Transforming the American Garden: 12 New Landscape Designs*, organized by Harvard University's Graduate School of Design. I immediately replied with an emphatic "YES!" I welcomed the opportunity to synthesize my philosophy of garden metaphysics and climatic design into a single artwork.

Refer to Figure 21.5, Garden of the Rose. The *Garden of the Rose* is contained within a Medieval-type triptych painted with Pompeian Red acrylic paint. Its protective glass evokes a sacred setting—a garden sanctuary and self-contained universe. The doors can be closed and the garden hidden and forgotten. When opened, it is a mystery to decipher. The garden plan represents a place of mystical contemplation and the different stages that one must pass through on the journey to enlightenment. In the centre of the predella is an oculus that reveals the hidden rose.

After working out my conceptual studies on tracing paper and deciding on a final layout, I transferred the drawing to 140 lb. hot-press Fabriano watercolour paper. I inked the drawing with a crow quill dip pen and painted it. When the watercolour had dried, I used a 1" bristle brush to brush the back of the drawing with Elmer's glue thinned with water. Holding my breath, I quickly pressed it down on the wooden frame and carefully smoothed it out until dry.

Further Visual Inventions: An Unlikely Mandala

Some time ago, after dining at a Chinese restaurant, the waiter brought to my table a take-home box. The empty white box caught my eye as it had no writing or advertising on it. I took out my fountain

21.5
Garden of the Rose. By Chip Sullivan.

pen and, on this blank canvas, sketched a view of a garden. I took home the illustrated box and kept it on a shelf. Years later, when brainstorming ideas for an exhibition, I spied the box, took it apart, and marvelled at its construction. I realized that its geometry presented a most interesting cubistic view of a garden design, somewhat like a mandala.

Refer to Figure 21.6, Almost Cubist: A Tea Garden in Golden Gate Park, San Francisco. This paper construction was drawn on a 10½ in square of two-ply, plate finish, Series 400 Bristol paper—an ideal surface for ink drawing. I outlined in pencil and then inked the lines with Faber-Castell Pitt pens in sizes XF and S. When complete, I scanned the drawing and had a Giclée print made on 140 lb. hot-press watercolour paper with archival black ink. I applied watercolour directly on the Giclée print. I initially folded the flat drawings into a "take-out garden" construction for an exhibition, but I found the unfolded drawings visually exciting and began this new series.

Sacred Woodland Spirits

In a recent exhibition titled *Flight of the Spirits: Artists Inspired by Miyazaki* at the Cukui Gallery in San Jose, California, I explored the mystical Camphor tree in Hayao Miyazaki's beloved animé film, *My Neighbour Totoro*. I created a sanctuary within a glass bell jar for the sacred tree in which the woodland spirit Totoro lives.

Refer to Figure 21.7, Totoro's Sacred Tree. After studying the tree shown in the film, I developed a series of thumbnail sketches and enlarged them to fit the scale of the bell jar. Using a light table, I transferred the sketches onto 140 lb. hot-press Arches watercolour paper. I refined the pencil layout lines and then inked them with a Faber-Castell Pitt Pen in size S. When the watercolour washes were dry, I cut out the individual pieces with an X-Acto #11 blade and glued them into place on the wood

The Spirit of Drawing 183

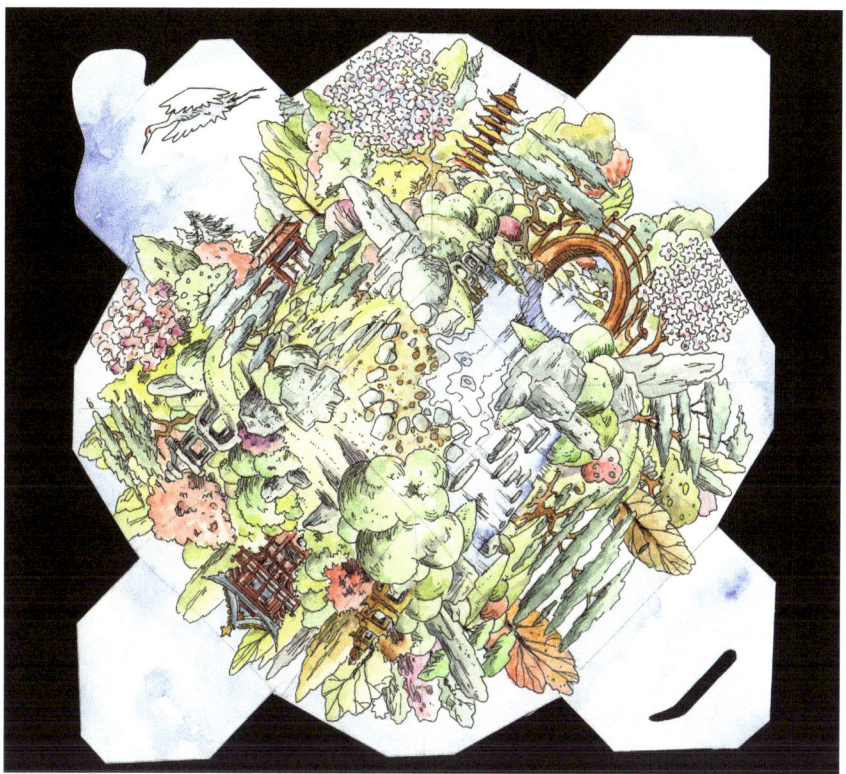

21.6
Almost Cubist: A Tea Garden in Golden Gate Park, San Francisco. By Chip Sullivan.

21.7
Totoro's Sacred Tree. By Chip Sullivan.

base with Elmer's glue. The front view shows Totoro's tree, and the reverse view reveals the portal of the tree spirits. Let us all try to find the hidden path to Totoro's tree.

When I wrote *Drawing the Landscape* many years ago, I wanted it to be more than just a technical instruction manual. I envisioned a textbook that would be inspirational and promote landscape drawing as an art form. I believed (and still do!) that drawing is an important tool for self-discovery and personal expression. I am humbled that the book has lived up to those ideals and exceeded my expectations. As I reflect on my decades of teaching at Berkeley, one of the gratifying rewards of teaching is to watch my former students cultivate those ideals in their careers as educators, practitioners, artists, and writers.[1] *Drawing the Landscape* is now in its fourth edition. I hope that my work and the ideas expressed in this chapter continue to inspire creative and imaginative approaches to landscape representation and art.

Note

1. I would like to acknowledge a few former students who have cultivated a successful career path in landscape architecture and visual communication under my mentorship. Jose Parral, a Rome Prize winner, is Professor and Chair of Architecture at Woodbury University in San Diego. His pedagogy similarly promotes the power of curiosity and the importance of learning how to "see." In her book, *By the Forces of Gravity*, Rebecca Fish Ewan—poet, cartoonist, and Professor of Landscape Architecture at Arizona State University—combines words and drawings as a powerful narrative. Landscape architect Patricia Algara employs techniques such as drawing on napkins for conceptual studies in the work of her award-winning firm BASE Landscape. Eddie Chau, former director of the UC Berkeley Extension landscape architecture certificate program, is an artist and landscape architect who teaches freehand drawing and landscape representation. His recent book *Random Imaginations* is a masterpiece of graphic invention. Film director, author, and production designer, Tom Hammock keeps his sketchbook by his side to determine camera positions and develop set designs. Currently senior lecturer at University of Huddersfield in England, Beatriz Gomes-Martin is writing her dissertation on the relationship between drawing and architecture. Seung Wan Hong, professor at Inha University, South Korea, incorporates freehand drawing projects into his architectural design studios. Shixian Shen, a visiting scholar from the National Academy of Fine Art in China, was inspired to again pick up his brush to record the life and scenery around him.

References

Guptill, A. *Color in Sketching and Rendering*. New York, NY: Reinhold Publishing Corporation, 1946.
Guptill, A. *Drawing in Pen and Ink and a Word Concerning the Brush*. New York, NY: Pencil Point Press, 1930.
Henri, R. *The Art Spirit*. Philadelphia, PA and New York, NY: J. P. Lippincott Company, 1960.
Hill, E. *The Language of Drawing*. Hoboken, NJ: Prentice Hall Inc., 1966.
Hubbard, H. and Kimball, T. *An Introduction to the Study of Landscape Architecture*. Boston, MA: Harvard Educational Trust, 1959.
Richmond, L. and Littlejohns, J. *The Technique of Watercolor Painting*. New York, NY: Pittman Publishing Corporation, 1952.
Shepherd, J.C. and Jellicoe, G.A. *Italian Gardens of the Renaissance*. Princeton, NJ: Princeton Architectural Press, 1986.
Slusky, J. and Sullivan, C. *The Impulse to Draw: Empowering Imagination for the Electronic Age*. San Francisco, CA: Norfolk Press, 2014.
Sullivan, C. *Garden and Climate*. New York, NY: McGraw-Hill, 2002.

22 Allegorical Drawings
Developing a Cultural Practice
Walter J. Hood

Landscape is perceived as a medium of a cultural practice. It offers agency to see and understand how race, class, environment, poverty, and discrimination are constructed and maintained in the U.S. landscape. The medium is a lens through which new interpretations and meaning can be constructed and critiqued. Landscapes are not neutral; they are multi-dimensional, shaped by their environmental, cultural, and political context. Landscape representation can be a key conduit for this agency, engaging the simultaneity of difference and function of landscape space. My interest in the double semiotic of a drawing, how I could decode the visual standards to validate culture, honour those unseen and unheard, and critique and celebrate simultaneously, was a product of my architectural education. Architecture historian Dianne Harris writes in explaining the cultural turn:

> Despite the fact that the 1980s became the decade to which post structuralism and cultural studies first impacted most fields within the humanities, their influence did not appear in landscape scholarship until the very end of that decade in the 1990s.[1]

My landscape architecture education favoured social factors through a behaviourist set of metrics. As an architectural student in the late 1980s and early 1990s, post-structuralism and cultural studies pervaded UC Berkeley's architectural pedagogy. Of particular influence were the narratives, drawings, and representations of Berkeley professors Lars Lerup, the *Pamphlet Architecture* series by Bill Stout, Steven Holl, and Mark Mack.[2] Both works engaged cultural studies in the representation of architecture and urbanism, and the *Pamphlet* series offered different models for architecture research and speculation, particularly #1 through #10 in the series.

Landscape representations are a form of storytelling. They are schemas by which to understand landscapes and their meaning. Whether simple hand sketches, drawn plans and sections, or perspective renderings, representations of a site are based on form or social, ecological, and physical production rules; in many cases, landscape representation can be normative and singular in its evocation of culture and then standardized through social factors and metrics. Social science research removes speculation and process from drawing representation, rendering culture as simply representing diverse populations engage in normative activities. In the last few years, the discussion has emerged around using "entourage" people supplied by digital platforms. Over time, through digital sharing, the same images of people are employed by designers, homogenizing culture, and difference. If not rendered dystopian for effect (ecological calamity), most landscape representation presents harmonic imagery and cultural satisfaction in how people are experiencing these landscapes.

The stories that I wanted to tell when I began my studio and research in the 1990s were the ones that were missing from landscape representation. These are the stories of Black people and the world in which they dwelled. The simultaneous articulation and disarticulation of the difference

DOI: 10.4324/9781003183402-22

between 'nature and convention' emerges from this dualistic view. The double semiotic as W.J.T. Mitchell references the "free space" or the informal spaces left vacant in the margins and bas-de-pages (bottom of the page) of medieval manuscripts inspired image and text collage.[3] Landscape drawings were produced against the backdrop of the cultural epoch of the 1990s, particularly, the rise in multiculturalism and the racial reckoning connected to police brutality. Different modes of representation create dual interpretations. Photography and narrative bring cultural and racial context to the forefront of landscape representation. Two sets of work from this period elucidate these ideas: *Jazz and Blues Landscapes* (1993) and *Urban Diaries* (1997). Both projects began as exhibitions, telling stories in three-dimensional space.

Blues and Jazz Landscapes

Inspired by narratives of user identification, their hobbies and work, and particular relationships among people common in 1980s speculative architectural projects, the representations are simple black and white line drawings. The absence of colour, as seen in Steven Holl's proposal for the N.Y. Highline in Pamphlet #7 and Mark Mack's 10 California Houses, Pamphlet #10, illuminated how narrative could foster architectural thought and speculation if the representation were non-hierarchical. The text is read as representation. By eliminating colour (particularly green) from the landscape drawings, they become neutral representations. Narrative and black and white photographs of people merge with plan and perspective and model (Figure 22.1).

Sixteen neighbourhood park designs are drawn in simple line plan and perspective drawings, alluding only to form and space. One construction detail reveals the materiality of a single element. A black and white photo portrait accompanies each plan and narrative. This is the setting for a set of design improvisations that put race and gender at the forefront in interpreting a design. The double semiotic, a traditional set of design drawings paired with image and narrative, forces the viewer to confront this new information. It is not a set of diagrams of uses and users; it is not behaviourist-inspired metrics; they are speculative scenarios that push the viewer to reimagine landscapes through the lens of the protagonist. The landscape is seen as the medium for the experience.

Narratives and fictions are gendered and multigenerational. The plan drawings are a transgressive attempt to decode/deconstruct the formal hegemony of landscape and its representation. Critiquing the cultural exclusion of difference and intentionally only using non-objective forms—that is, circle, square, and triangle—situates the narratives as formal instigators. As a result, new forms emerge with little to no meaning until connected to the narrative and the image of the inhibitor. With these, the plans are simple abstract drawings.

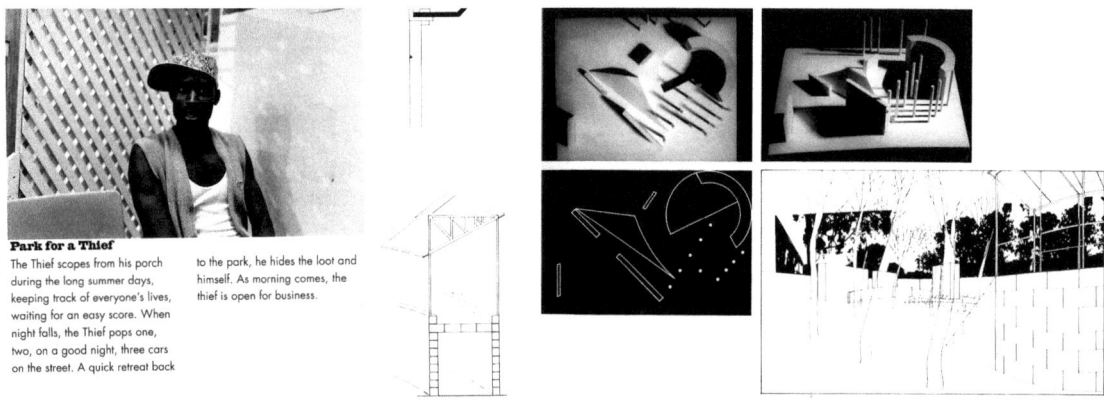

22.1
A Park for a Thief. Black and white photographs of people merge with plan and perspective, and model, from the Blues and Jazz collection.

Urban Diaries

Lars Lerup's drawings from the 1980s and 1990s are design actions. They are a rejection of behaviourism. Lerup offered a particular critique through drawing and representation expression; like the abstract painters of the 1980s, the erasure was present in the drawing, noticeable through the use of white-out fluid and scratching out with a pen. Colour is used to emphasize to particular architectural elements, like a red handrail. Lerup's drawings elucidate a cultural shift, a schema for difference, a schema for correction. Using the axonometric, partial perspective, and long black shadow projections, the drawing reads like a graphic novel.

If *Blues and Jazz Landscapes* brought people to the forefront of design representation, *Urban Diaries* attempt to merge normative representation into a novel-like, semi-fictitious world through multi-layering. Each representation suggests a singular time, the multi-dimensional aspects of daily life orchestrated as a daily diary. Each day, change occurs in a space based on its inhabitants. Akin to a play, the proscenium/backdrop changes are underscored by the similar non-objective formal set of organization. The representations are layered—text on acrylic, sketches from tracing paper collaged on the drawing, photos of people, plans, sections, parallel line projects—to tell a different story of landscape.

There are multiple expressions of drawings in *Urban Diaries*. The park-space drawings read like an open book as the drawings are horizontal, merging plan, axonometric, and section. At the bottom of the page are painted diagrams and collaged sketches from my sketchbook. The drawing's centre is a black and white photo; plan, section, and axonometric drawings sit on an airbrushed backdrop ground of colour (Figures 22.2–22.6). The large freeway drawings emerge from a gesso ground on a black and white aerial, featuring the context that the freeway destroyed (Figure 22.7a and 22.7b). The black and white photos merge with the aerial as colour brings the design idea forward. And last, the landscape objects are exploded axonometric on an airbrushed background (Figure 22.8). The different drawing formats are connected through the colour grounds and photographs.

Landscape representations from both projects offered different semiotic means to discuss tough subjects. Conversations arose around gender identity in these speculative works. Normative behaviour was replaced with counter-cultural acts and defiance. There was House for a Prostitute, where the shadow from the church appears daily in the living room; a Park for a Thief (Figure 22.1), where the image of a young black guy with his hat turned backward; a Drive-through Brothel (Figure 22.8) that makes sex acts visible; and stories of liquor stores, freeways, yard sales, and street stadiums that bring the everyday from marginal communities to life. Looking back at this work, it seems as significant today as it did 20 years ago.

22.2
25th Street. Mixed-media, black and white photographs, airbrushed, pen and ink, from the Urban Diaries collection.

22.3
29th Street Houses. Mixed media, from the Urban Diaries collection.

22.4
Durant Park Recycling. Mixed media, from the Urban Diaries collection.

Allegorical Drawings 189

22.5
Cooking at the Park. Mixed media, from the Urban Diaries collection.

22.6
Neighbourhood Watch. Sketch Model and Painting, pastel, from the Urban Diaries collection.

22.7
(a) and (b) Church Overpass. Black and white aerial of the site, overlaid with colour plan collage (trace) and black and white photographs, landscape objects are exploded axonometric on an airbrushed ground, from the Urban Diaries collection.

22.8
Drive-through Brothel. Airbrush colour, from the Urban Diaries collection.

Notes

1. Harris, D. "The Postmodernization of Landscape: A Critical Historiography." *Journal of the Society of Architectural Historians*, 1999, 58(3): 434–443, p. 436.
2. Holl, S. *Pamphlet Architecture #1–10*. Princeton, NJ: Princeton Architectural Press, 1998.
3. Mitchell, W.J.T. *Landscape and Power*. 2nd ed. Chicago, IL: The University of Chicago Press, 2002.

References

Harris, D. "The Postmodernization of Landscape: A Critical Historiography." *Journal of the Society of Architectural Historians*, 58(3) (1999): 434–443, p. 436.
Holl, S. *Pamphlet Architecture #1–10*. Princeton, NJ: Princeton Architectural Press, 1998.
Mitchell, W.J.T. *Landscape and Power*. 2nd ed. Chicago, IL: The University of Chicago Press, 2002.

23 ASPECTS [of] Design Representation

ASPECT Studios[1] and Jillian Walliss

For ASPECT Studios, landscape representation is integral to design processes. Consequently, there is no identifiable graphic 'style' for this multi-studio practice, which instead considers representation as a suite of critical 'techniques' for exploring and generating design outcomes. This position can be traced to the practice's origins and to the major generational changes in landscape architecture in mid-1990s Melbourne. Many of the now studio leaders are graduates of the RMIT University landscape program, which was internationally influential in advancing the role of landscape representation from a passive image of the designer's predetermined vision (constructed in sketchbooks, the back of napkins, on a sheet of butter paper) to part of more powerful generative processes, spanning critical, and creative thinking.

From its original Melbourne studio, ASPECT Studios has expanded to nine studios in Australia, China, Vietnam, and the Middle East. The spread of representational approaches evident in these offices reflects the evolving challenges of a contemporary design practice embedded in the Asia/Pacific region. Increasingly, ASPECT Studios is known for its civic work, which spans from large-scale infrastructural projects to urban design and cultural precincts. This focus has generated a repertoire of techniques that move fluidly across the digital and analogue to engage with the technical complexities demanded of large-scale, multi-disciplinary projects, while also exploring human-scale experiences and a strong cultural engagement.

Examination of the Melbourne, Shanghai, and Sydney studios reveals particular representational agendas shaped by the specific design challenges and opportunities that emerge from their unique geographic and cultural positions. For example, the Melbourne studio has emerged as an industry leader in digital design technology and is considered an early adopter of 3D modelling, parametric modelling, and BIM implementation in Australia. This has been driven by a 15-year engagement with large-scale infrastructure projects, which require alignment and integration with engineering and architecture practices. In contrast, through being embedded in the cultural and ecological complexities of the contemporary Chinese city, the Shanghai office works closely with diagramming as a technique for developing spatial and programmatic relationships between culture, recreation, commerce, and ecology in dense urban typologies. For the Sydney office, the 3D view explored through hybrids of the hand sketch and the digitally produced perspective generates the contextual responses required to insert new civic spaces into Sydney's rich urban environment. While individual designers' interests and skills certainly influence these differing representational techniques, these representational modes of working are equally selected for their ability to address particular design challenges.

The Dynamic Digital Model

For the Melbourne studio, the parametric digital model has proved a powerful medium for dynamically bringing alive the potential of landscape across a range of scales. From large-scale infrastructure collaborations with engineers and architects through to more sculptural and detail-driven explorations with artists, the digital model has allowed studio director Kirsten Bauer to work fluidly across form-finding and making. The responsive qualities of the digital model have been particularly valuable when working with expansive landscapes.

This was well demonstrated in their involvement in the Victorian Desalination Plant and Ecological Reserve (2012), where a responsive parametric model developed in Maya® provided the necessary computational power required to participate in such a fast-paced, multi-disciplinary collaboration. In contrast to a static physical model, the parametric model with its dynamic feedback allowed ASPECT to understand and explore the technicality and scale of their proposed monumental landforms bordering the plant and its green roof. Equally and significantly, it facilitated the simultaneous testing of the experience of these landforms from strategic vantage points. The screenshot shown in Figure 23.1 illustrates how this parametric model breaks from a more conventional linear design process of concept, representation, and documentation, instead towards compressed methods of design generation and documentation. With an ability to input precise data and real-time feedback, the digital model allows the landscape architect to work creatively and accurately, and, most importantly, quickly.

This experience of digital terrain modelling has been explored more recently in the Middle East, this time proving essential for understanding how the manipulation of the subtle but expansive desert landscape is experienced from the perspective of a moving car. Figure 23.2 offers the client a sense of scale and space which would be impossible to communicate through plans and sections.

At the civic scale of Perth's Yagan Square (2019), the potentials of the digital model were explored further as a medium for poetic collaboration with artist John Tarry. This new square was underpinned by an innovative, creative template developed in collaboration with the Whadjuk people (the traditional owners of the land), which guided the designers and artists in the development of the public realm. Yagan Square's location on a former ephemeral wetland inspired a series of connected water experiences known as the 'waterline.'

Working with Rhino®, Revit®, and Grasshopper®, ASPECT transformed a digital expressive 3D sculpture of an 'egg-like rock' developed by the artist into a digital, detailed parametric model. This

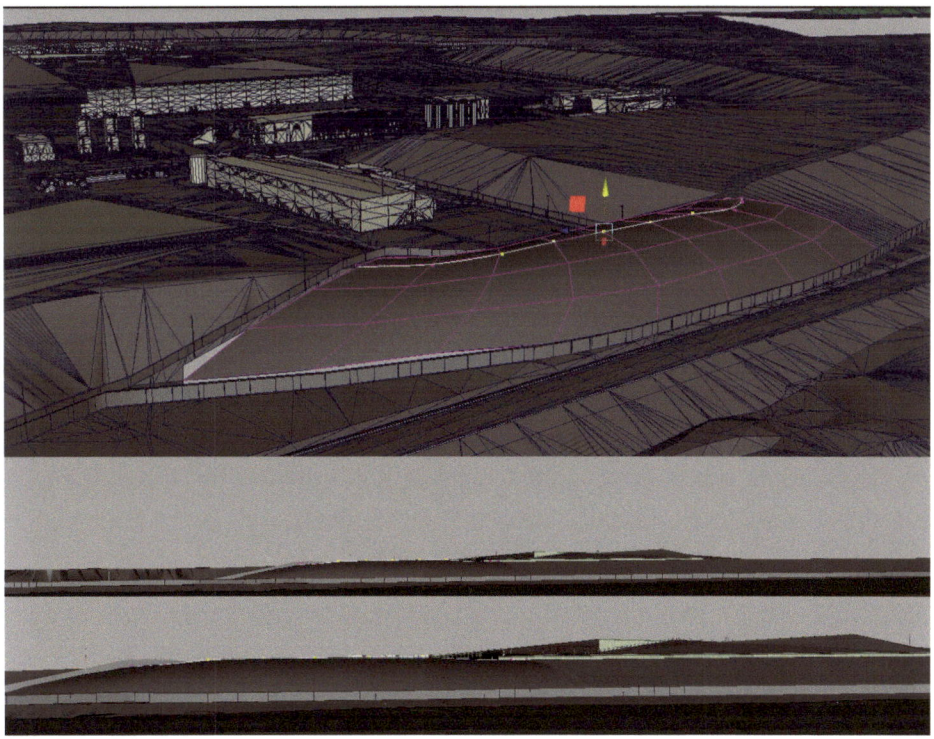

23.1
Screenshot of the Maya® model used for the Victorian Desalination Plant and Ecological Reserve. The model could be manipulated in real time to test the effect of the landforms.

23.2
A rendered image constructed from a Rhino® model which highlights the experience of moving through the expansive desert landscape of the Middle East in a car.

rock—as both a virtual sculpturing tool and an actual proposed form—informed the design of the critical spatial moment where the water first bubbles up from the ground before cascading down a series of carved granite channels through the square. As shown in Figure 23.3, the digital rock was used to simulate a carving process into the ground plane and generated the final form. Further, the model was used to simulate the 'wetness' of the scheme, valuable in interrogating the relationship between water flow and the proposed contours. The precision of this form was further developed using a Revit® model, which can also guide the carving of the physical granite stone through CNC milling. As Bauer highlights, these digital models, therefore, operate as a powerful representational technique for finding form and, ultimately, making form.

The Sketch Diagram

The diagram forms a particularly valuable representational technique for the Shanghai office. Diagrams, along with scribbles and sketches, form essential tools for the Studio Director Stephen Buckle, who is dyslexic, to communicate critical and complex information. Further, in the context of living and designing within cultures where English is not a first language, and often not spoken by some clients, diagrams are essential for communicating design ambitions and ideas through all design stages. The diagram, Buckle observes, can communicate the 'why' and the 'how' of an idea, issue, or solution. For designers, diagrams used as part of the design process offer a powerful conceptual device for reducing complexities into a relational language that can inform design possibilities.

Diagrams, especially on large projects, act as an important device for orientating the design team. With the ability to clearly describe the many layers and subtleties of the design process, diagrams help everyone understand a common vision and direct them towards shared goals in the design development process without confusing narratives and concepts.

Their value is well demonstrated in the design of the Hyper Lane development in Chengdu, Sichuan (2020). Almost 2.4 km long, this multi-level linear sky park offers a mix of cultural and lifestyle programs, along with facilitating critical links to local transportation nodes and the arts and music university campus. The diagram forms a crucial tool for organizing and conceiving connections and programs across these multiple levels. The abstraction and editing required of the diagram distils information into a conceptual idea for engaging with the cultural, economic, and ecological complexity of dense urban typologies, which can be easily communicated with clients and design teams.

Figure 23.4 highlights the diagrammatic sequence used to succinctly communicate the notion of 'giving form to the idea of community' for The Urban Gallery (Phase 1 of the Hyper Lane). This is

ASPECTS [of] Design Representation

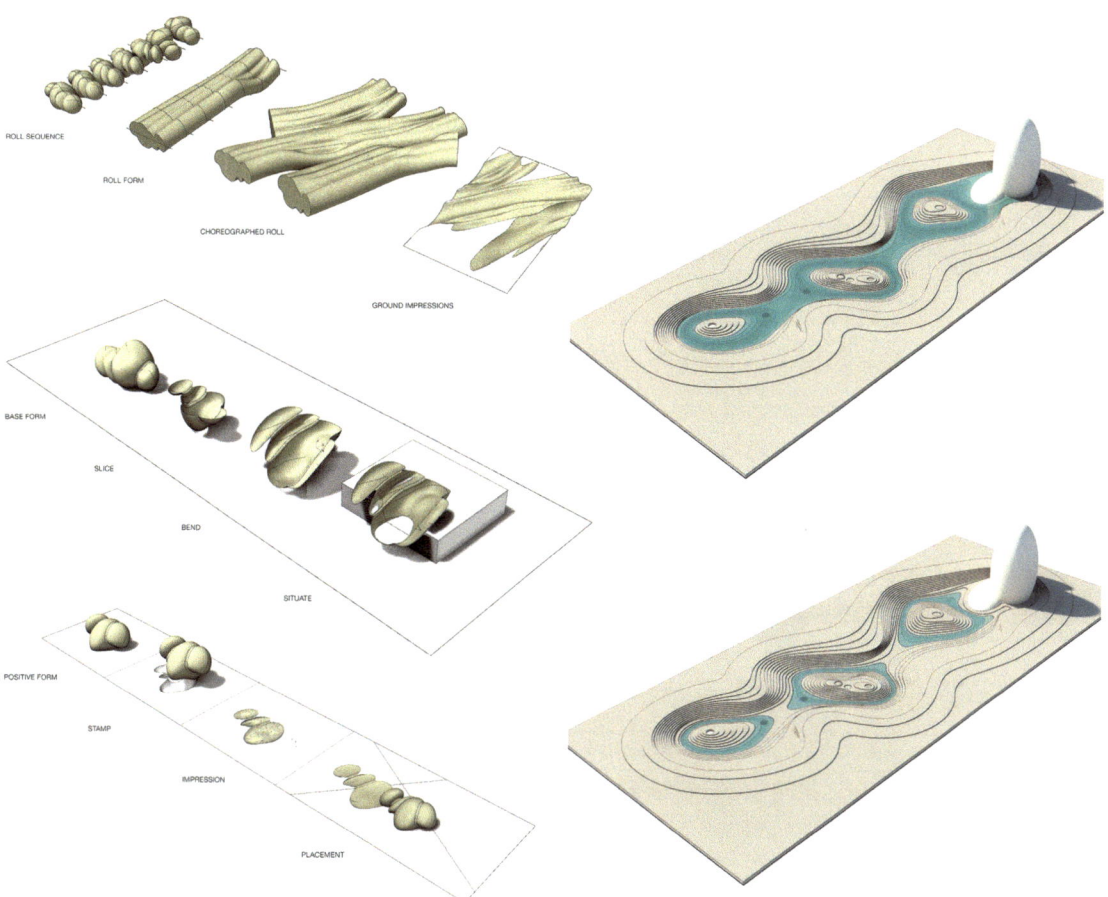

23.3
Digital design explorations featuring a digital 'rock' used to simulate a carving process and generate form.

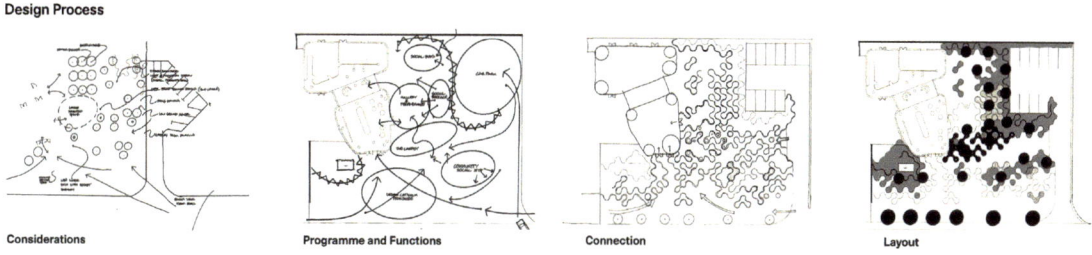

23.4
A series of diagrams of the Urban Gallery used to communicate early-stage site considerations, programme, intent, and design layout.

expressed graphically to inform the direction of a vibrant youth-oriented point of identity that reflects the artistic and creative culture of the district.

Shifting into a more detailed interrogation of form and urban detail, the diagram becomes a valuable representational tool for exploring and expressing how a singular unifying design intent can adapt, morph, transform, and respond to the varying functions, systems, and experiences of place. The diagram shown in Figure 23.5 highlights how a singular gesture is to be extruded and reinterpreted into form and respond to an interlocking social program. Sustainable stormwater management, social pockets, terraced platforms, permanent elements, curatable features, performance spaces, and intimate work pods are connected and unified by the diagrammatic clarity of a single line. The diagram, therefore, has a strong relationship with the plan. One will inform the other. As the diagram

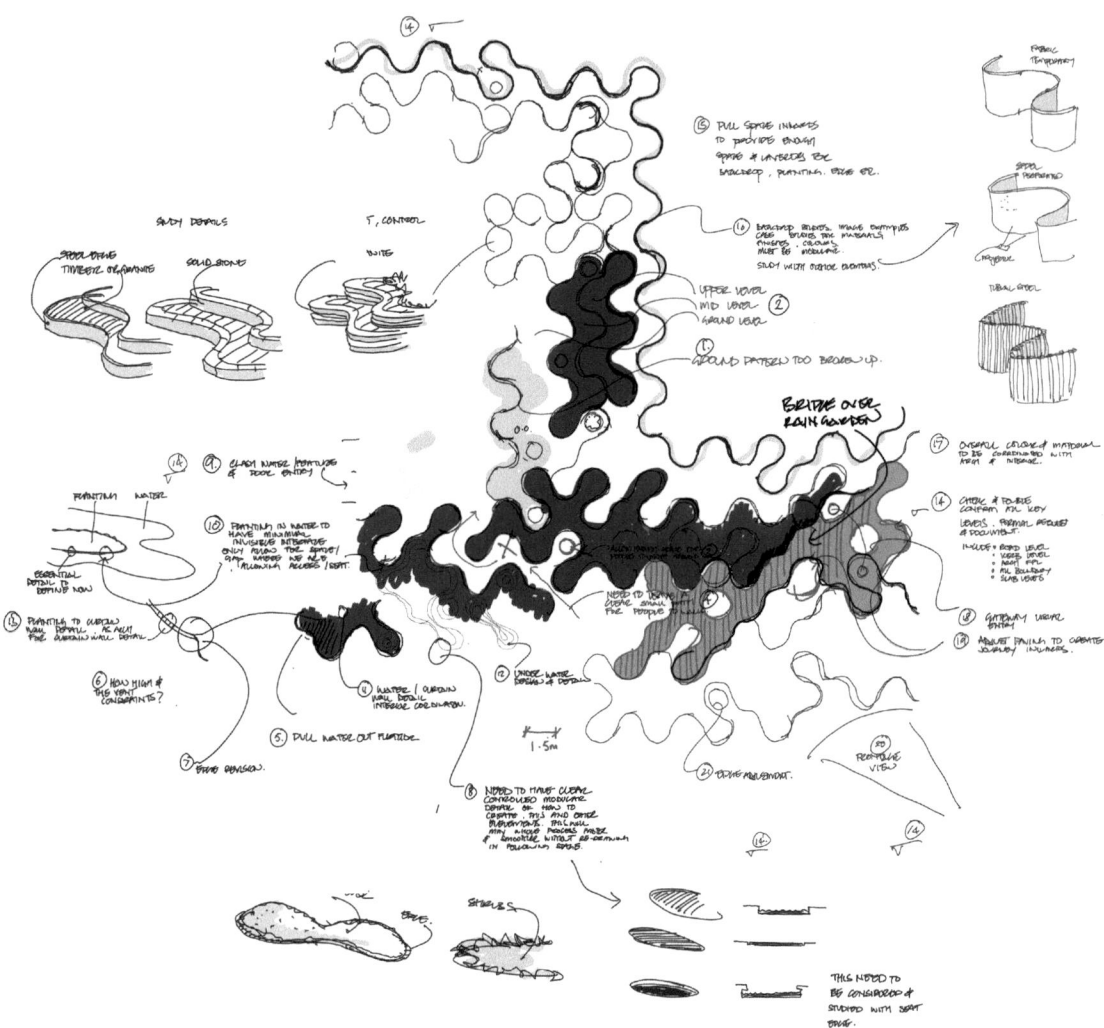

23.5
Diagram demonstrating how the site concept for the Urban Gallery could be translated into form and program.

is developed, it is transformed into a project site plan which offers a resolution of the multiple layers explored and expressed in the diagram.

The Drawing

Turning to the ASPECT's Sydney office, the combination of Sydney's sandstone topography, harbour, and post-industrial sites drives a robust contextual design approach for inserting new civic spaces into this complex and often irregular urban fabric. Working within this realm, the 3D image is essential for exploring and communicating a human-scale experience of proposed spatial sequences. Examples range from the quick hand sketch, which succinctly abstracts critical information, form, and phenomena, through to more developed perspectival views. For director Sacha Coles, the beauty of a rough and emotive sketch brings everyone, including clients, into a conversation of possibilities. The unfinished nature of the quick sketch, states Coles, permits the addition of more layers, leaving room for new ideas while giving a clear direction. The drawing highlights creative and complex ideas in a simple way, with an economy of effort expressed through line work.

The quick sketch was pivotal in developing The Goods Line Project (2015), which transformed a forgotten inner-city urban space into a new connected civic domain. The hand-drawn sketch shown in Figure 23.6 establishes key spatial sequences for this new public spine. These types of sketches used in the early stages of design indicate approaches to the project without offering a complete

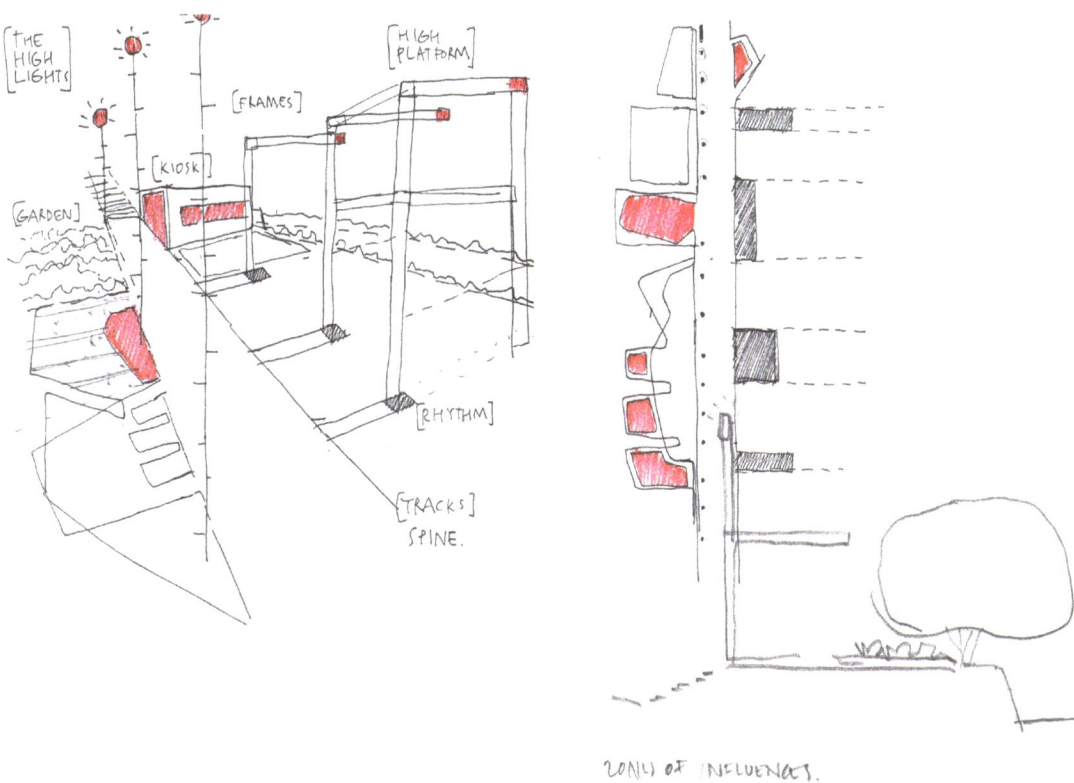

23.6
Early hand-drawn concept sketch for the Goods Line highlighting key spatial experiences and sequences.

23.7
A more developed composite concept image of the Goods Line which emphasizes the experience of being in the space.

design—which is an essential tactic for success in design competitions. Therefore, fast sketches aim to draw out and exaggerate—almost caricature—experience so that it is evocative rather than literal or to scale.

The more developed perspective of the Goods Line shown in Figure 23.7 presents a stronger sense of the project's identity. Aspects are heightened to develop a more meaningful image, which

23.8
Composite concept image of the Ian Potter Children's Wild Play Garden. The image was constructed from the viewpoint of a child.

conveys a sense of what it is like to inhabit the space. Often, these more developed images originate from a sketch, with digital tools such as the iPad-focused Procreate® package facilitating the hybridization of media, bringing together drawings, models, and all sorts of other media. Importantly, for The Goods Line Project, this representation includes the contextual edges bordering the proposed space as it threads its way through the existing urban fabric.

Foregrounding a feeling of immersion was central to the design for the Ian Potter Children's Wild Play Garden (2017), which is located in one of Sydney's oldest parks. ASPECT's scheme drew on the existing site features of fig trees and parkland and a simple palette of natural materials and elements to develop an experiential environment for children to explore, experiment, socialize, and learn through play. This required the designers to inhabit the mind of the child, a view that was explored through drawing conventions where some experiences are drawn bigger than others, and certain details, textures, or plants are highly exaggerated (Figure 23.8). Playing with ideas of scale, texture, and experience in the drawing introduces emotion into the design, revealing how a child might encounter mysterious spaces or how a space might provoke the imagination. Therefore, the drawings developed for Wild Play were creative, imaginative, and loose, with a childlike and naïve expression. For other projects such as the Goods Line, the representation involved far more crisp lines that reflect the post-industrial project's qualities more directly.

Representation Is Design

ASPECT Studios demonstrates the integral relationship that design representation plays in the design process. They are not two separate things, but rather, are co-dependent. Each studio is encouraged to find and invent techniques which best suit the challenges of the project, the client, and the site. This identification or search for the appropriate representation technique is critical to developing a design position and ultimately influences the final design outcome. Representation, therefore, is not a passive 'illustrative' part of a design process; it is highly influential in establishing design direction. Working in the fast-paced world of contemporary landscape architecture, speed becomes an essential part of the representational discussion. Whereas a century ago, designers may have conceived of parks with a few hand sketches and a plan, contemporary design projects are procured, designed, and documented at a rapid rate. Therefore, every studio must work across multiple techniques, fluidly moving between the analogue and the digital, the quick and the precise, and the poetic and the technical. For ASPECT Studios, landscape representation is not a question of visualization; rather, it is a driver of design.

Note

1 Text is based on observations from the three Studio Directors Kirsten Bauer (Melbourne), SachaColes (Sydney), and Stephen Buckle (Shanghai). All images are copyright to ASPECT Studios.

24 Every Picture Tells a Story
The Iconography of GROSS.MAX. Imagery

Eelco Hooftman

> You must understand that a landscape plan can neither be imagined, sketched, drawn, coloured, or retouched, by any but a landscape painter.
> —René Louis de Giradin, *De la Composition des Paysages*, 1777[1]

The pixel is our pigment, the computer screen our canvas. Our inspiration comes from a myriad of sources. We certainly proclaim landscape architecture to be a creative discipline closely related to the arts. Our ambition is to explore landscape as an expanding sensory field. The images we create act as an eye-opener towards a new landscape. To provoke a "change of scenery" is a key motive behind our works. The landscape is part of visual culture. Sometimes, we long for those days when landscape was to be considered not just important—but rather the most important branch of aesthetics. In this writing, we do not dwell on technique but rather on poetry. We collected various visual materials—rather like magpies—to clarify and illustrate particular points.

The word 'landscape' is derived from landscape painting. The painter of the landscape was called a 'landscapist'—we at GROSS.MAX. rather like that word, 'landscapist'; it has a good feel to it, a representation of the landscape without all the ballast of the added architecture. The word landscapist balances a subtle hint of the artist with the allure of the escapist. Of course, in our work, we study the old masters not only regarding composition but also, especially, the use of light. For us, light and atmosphere are the essences of good landscape visualization. We are inspired by Titian's atmospheric landscapes, created by a painting technique of semi-translucent, multi-coloured layers of pigment, brilliantly catching sunlight glinting through the foliage. We are inspired by the biblical floodgate of natural light streaming through the vast church interiors, as painted by Pieter Saenredam. We are influenced by Jacob Van Ruisdael's paintings that capture the 'essence' of the Dutch landscape, depicting a low horizon and pregnant clouds with beams of sunlight which draw the viewer, like searchlights, into the epicentre of the composition. The golden glow, magic wand brushstrokes of Claude Lorrain in his paintings, and the immense sky with floating clouds drifting by as seen through the eyes of John Constant influence us. Maybe, with the revival of the 'landscapist,' we should pump up the volume of that quintessential British notion of the picturesque.

As landscape architects, we are oscillating between the real and the ideal. Our images do not exist in their own right but as a means, a missing link, in the process of making something which is not just representational, but 'transformational.' For many architects, the plan and diagram are the generators of designs. Visuals are outsourced to professional rendering companies at the very tail end of the design process. The photo-realistic images are used to convince the jury, seduce the client, and to impress the end-user. In general, they look amazing, but they seem remarkably similar and uniform, as if you are presenting a hyper-realistic freak show of ready-made, easily digestible, and lookalike architecture. The same blue sky, the same cloned trees, and sometimes even the same people are all vaguely familiar, like a virtual tribe of long-distance relatives. Such visuals which aspire to photo-realism create a closed world, instead of letting people see for themselves with their imagination. On

DOI: 10.4324/9781003183402-24

the contrary, many of the images we produce originate at the early stages of the design process. They do not show the comparison between before and after, but rather the state of the 'in-between.' Each viewpoint represents a point of view. Images are not only representation but also speculation and an expression of experimentation. Each of our images is created to elicit a particular emotional response. The art historian E.H. Gombrich stated that "What a painter inquires into is not the nature of the physical world but the nature of our reactions to it."[2] Similarly, one could argue that what our landscape images inquire into is not the nature of the constructed project, but our intentions.

In his 1821 *Confessions of an English Opium-Eater*, Thomas de Quincey described it as a condition of entrancing dreams—especially those "waking-dreams" occurring as one is half-asleep and half-awake when prosaic reality mixes in strange and inspiring ways with memory and imagination. Landscape, in our vision, is such deliberately induced phantasmagoria that unfolds in the half-asleep and half-awake condition between concept and context in the twilight zone between imagination and realization.

Our early photomontage images have been compared to that of surreal collage. Max Ernst defined collage as "the systematic exploitation of the coincidental or artificially induced encounter of two or more unrelated realities on an apparently inappropriate plane—and the spark of poetry that leaps across the gaps between them."[3] This 'spark of poetry' caused by dialectic juxtaposition is not only a visual language but also often alludes to a narration, often with literary associations. Every picture tells a story; the medium is both the message and the massage. Image and idea are intertwined. The collage technique in itself also represents the condition of the contemporary post-modern landscape, what we may describe as a montage-landscape, a hybrid assemblage built from fragments and layers which at its best read as a palimpsest of time.

Equally important is the influence of pop art, such as the flower paintings by Andy Warhol or the collages by Richard Hamilton. Also, we are intrigued by the drawings of Archigram and Super Studio. All of this resonates with the image production of popular advertising art, which teaches us how to organize optical qualities to communicate messages. Photographic montage, contrast, colour saturation, transparency all applied to engage with an audience on the level of visual experience. Landscape is about movement through space and time. We deal with duration in serial and sequence. Like the world of painting, we at GROSS.MAX. are equally fascinated and inspired by cinema. As landscape architects, we have much in common with filmmakers. Our heroes are directors the likes of Michelangelo Antonioni, Wim Wenders, Andrey Tarkovsky, and Werner Herzog.

Our images are produced in the studio workshop. It is difficult to attribute them to specific authorship. Undoubtedly, a connoisseur could distinguish certain characteristics, and for sure, some of our most striking images were conceived with the late Ross Ballard, who died far too early in 2017.

I would like to share a collection of our drawings that capture our style and visual experience. The Hofplein Viaduct proposal was the Highline 'avant la letter' (Figure 24.1). Our proposal incorporated a glasshouse as a hedonistic pleasure zone superimposed on the artefacts of the old railway station. It envisaged encapsulating an organic orgy of orchids amidst other extravagant, exotic plants. Anarchic, amorphous, and monstrous—landscape architecture as skilful, accurate, a magnificent interplay of assembled vegetation under the light. The solitary person standing on a rocky outcrop is, of course, the iconic ca. 1818 *Wanderer Above the Sea Fog* by the German Romantic landscape artist Caspar Friedrich. Maybe he is—who could possibly tell?—contemplating the 1940s bombardment by his fellow—less romantic—countrymen who destroyed this part of Rotterdam. The air laden with the perfume of plants intoxicates the figure in the foreground. As André Breton wrote, "Everything is dependence on our deliberate hallucinations."[4] Years after we made the image, when in the Museum of Fine Arts in Boston, I was shocked when I saw one of the orchid paintings by Martin Johnson Heade, *Cattleya Orchid and Three Hummingbirds* (Figure 24.2). For a moment, I was transfixed, drawn into the painting as by an invisible rope entangled by the dense foliage, mossy branches, and scattered sun rays penetrating through the hazy moisture-laden mist highlighting the sensuous erotic charged *Cattleya labiata* orchid. I could feel the hypnotic buzz of hummingbirds and the fragrant musty smell. I was struck by this feeling. As contrast, the close-up flower was juxtaposed against the wide-angled landscape, the use of dramatic sky, and the slightly oversaturated colours. George Bataille wrote in

24.1
AIR Rotterdam. By GROSS.MAX. / Ross Ballard, 2001.

24.2
Martin Johnson Heade painting-oil on wood. Cattleya Orchid and Three Hummingbirds, 1871. Credit: Gift of the Morris and Gwendolyn Cafritz Foundation, National Gallery of Art.

The Language of Flowers regarding orchids: "Plants so shady that one is tempted to attribute to them the most troubling of human perversions."⁵

Refer to Figure 24.3. Normally, architecture destroys landscape, and the rarity of demolishing a building to create a landscape was quite empowering. The project is for a temporary garden at Strathclyde University, Glasgow. We kept remnants of the former Victorian hospital as follies in the landscape and recycled its sandstone into gabion retaining walls. The world of advertising inspired us to create this landscape visualization. The Chair of Strathclyde University had instructed us to

24.3
Rottenrow Garden. By GROSS.MAX. / Kenton Wilson, 2003.

24.4
New Town/New Nature. By GROSS.MAX. / Ross Ballard, 2003.

remove the girl, as deemed overtly sexy. In our final image, we replaced the girl with a spotty, drippy boy with red, greasy hair. Apparently, this was fine and compliant with the image the university wanted to project to the students it wanted to attract. Glad they kept the little boy with the fox on his back!

Refer to Figure 24.4. Cumbernauld New Town outside Glasgow, Scotland, will one day be rediscovered as an ancient twentieth-century temple dedicated to worship the Motorcar. Archaeologists of the future will be puzzled by its remains. It is one of the largest concrete structures in Western Europe, a kind of drive-in Stonehenge. A town devoted to the worship of mobility based on strict fundamentalist segregation of human and motor. A compact city with a mega-structure, multi-level, single-envelop town centre. A rare crossbreed of Italian Hilltop Town and American Shopping Mall. Long before the now compulsory inner-city green-wash, we revisited Cumbernauld New Town with an early connection to New Nature. Ecology is seen as software downloaded on the hardware of the city. Figure 24.4 is an overlay of images including an original black and white visual of the first

24.5
Global Warming/Local Freezing. By GROSS.MAX. / Ross Ballard, 2005.

24.6
Garden for a Plant Collector. By GROSS.MAX. / Ross Ballard, 2005.

phase of Cumbernauld New Town as the backdrop. The original, with its flashy Cadillac convertible, depicts a fantastic hint of the American Dream, and we coloured it fuchsia pink to catch the eye. Its signed number plate gives away the date of our image.

Refer to Figure 24.5. To counterattack the effects of global warming, the image suggests the concept of local freezing and 'climate interchange' by using Torness Power Station on Scotland's East coast like a giant cool box to create a nuclear-powered iceberg. The heat generated to cool the iceberg is used to make hot water lagoons in the nearby cement quarry, made a dazzling blue by the limestone rock formation. The site is near Siccar Point, where James Hutton, an eighteenth-century Scottish geologist, discovered in 1785 in the folded geology of the sea cliff the visual evidence for his theory on 'unconformity.' In his theory, Hutton described the Earth as a system without beginning or end, constantly changing, yielding to the powers of wind and water, heat and cold.... The panoramic image—arguably GROSS.MAX.'s most iconic—captures the atmosphere of climate change as the new sublime. Little hints such as the girl's pink dress picking up the pink of the flock of flamingos overhead, the atmosphere of steam, and clouds set against the dark moisture saturated sky reflect the masters' stroke talent of the late Ross Ballard. At his funeral, Ross left us in a coffin imprinted with the iceberg image and accompanied by David Bowie's soundtrack of Space Oddity.

Refer to Figure 24.6. The Garden for a Plant Collector is situated adjacent to Charles Rennie Mackintosh House for an Art Lover at Bellahouston Park, Glasgow. The glasshouse contains an

24.7
Vertical Garden. By GROSS.MAX. / Ryuichiro Noda, 2006.

exotic tropical plant collection set in a mysterious temperate woodland. Our original inspiration was a chapter in the novel *Against Nature* written in 1884 by the decadent author J.K. Huysmans. The protagonist of this novel, Des Eissentes, is obsessed with natural flowers which look artificial. The Cartesian precision of the architecture is in contrast to the overgrown vegetation. Each compartment of the glasshouse is differentiated by its distinct tone of tinted glass that vividly sets alight the image against the dark monochrome woodland setting. Careful 'Peeping Tom' observers of the image could distinguish Le Corbusier and Mies van der Rohe playing a game of hide and seek under cover of the dense, voluptuous foliage.

Refer to Figure 24.7. The vertical garden was a collaboration with the artist Mark Dion. While Mark explored the project with his characteristic red/blue pencil drawings, our visualization with its view from above accentuates the spatial expanse of lines, planes, and shape of the red-welded steel frame overgrown by a lush, rampant foliage of vegetation. The image 'pulls' the viewer by optical leverage or vortex down into the structure and plug hole vanishing point. The picture plane dissolves into the overlapping interpenetration of foreshortening planes and the open skeleton of lines. The amplified perspective, angle of vision, and its extreme distortion—rather than static linear perspective—was applied to inject the picture space with its dynamic expression, vitality, and a particular sensation of vertigo. With the benefit of hindsight, reference would be to Piranesi, the short story "Descent into the Maelstrom" by Edgar Allan Poe, and constructivist composition by the likes of Rodchenko and Malevich. Certainly, as seen from above, the figures below are reminiscent of early Bauhaus photographs by Moholy Nagy.

Refer to Figure 24.8. For the Northern City exhibition, we made a pictorial survey of Edinburgh, presented in a red leather-bound booklet animated with a series of panoramic images as a mixture of old engravings and contemporary overlays. Both the book cover and the composite images are reminiscent of the famous Red Books by the eighteenth-century landscape architect Humphry Repton. The physical aspects of Edinburgh, such as coast, valleys, and hills, form an important reference as representing 'the long time span' set against the 'delirious experience' of the city in a state of flux. The combination of images and quotations diagnoses Edinburgh as a strange case of split duality. The image selected depicts a suburban fox hunt across the Edinburgh Bypass, as foxes these days no longer live in the countryside but have become scavengers of suburbia instead. Of course, the commercial signs and advertisements are a slight nod to Venturi's "Learning from Las Vegas."

Every Picture Tells a Story 205

24.8
Old Town New Town No Town. By GROSS.MAX. / Richard Beven, 2006.

24.9
Gleisdreieck. By GROSS.MAX. / Ross Ballard, 2006.

Refer to Figure 24.9. You could imagine this image as a still from a movie. The setting is Berlin, just south of the Landwehr Canal, where a former railway yard is designated to become a new urban park. Instead of showing the park, we show the urban edge with glazed transparent structures providing access to the elevated park on higher ground. We are reminded of *One-Way Street* by Walter Benjamin or *Himmel Uber Berlin* by Wim Wenders. The ink-black night sky, the opaque transparency of the glazed structures, the glow of scattered illumination, the blue-toned girl in the foreground all amount to a sense of suspense. The image has grit and an understanding of the sublime. Like Picasso, we at GROSS.MAX. went through our own blue period.

Refer to Figure 24.10. "Who is afraid of the urban void?" was the rhetorical question we asked ourselves when designing the former Tempelhof Airport and transforming it into a new generation of

24.10
Tempelhof Airport. By GROSS.MAX. / Nina Maley, 2011.

24.11
Xi'an Master Garden. By GROSS.MAX. / Lois Farningham, 2011.

Volkspark for the twenty-first century. We envisaged the park as a prairie for the contemporary urban cowboy. Nietzsche wrote in *The Joyous Science*:

> At some point, probably soon, we will realize what is lacking most of our big cities: quiet, open, expansive places for contemplation, places with long, spacious passages for bad or hot weather, where no sound of carriages or crier is heard and where a finer sense of propriety would even forbid a priest to pray aloud: building and complexes that as a whole express the sublimity of contemplation and withdrawal.[6]

The image depicts the wide expanse of horizon and the perspective of the runaway runway.

Refer to Figure 24.11. In a select design completion invitation, we were invited to design a Master Garden at the Xi'an horticultural expo 2011. Our garden was a tribute to the English plant hunter Henry Wilson, who in the early twentieth century, like a hyperactive squirrel, collected thousands of new plant specimens from across China to feed the exotic plant addiction of both British and American garden junkies. We created a circular shrine built of stacked terracotta pantiles. We used the original engraving of various plants sourced from old catalogues and plant books. The boy we found in an old photo album; his position off centre creates a dynamic feel to the otherwise straight frontal composition.

Refer to Figure 24.12. There is a close connection between the romantic and picturesque English Landscape style and the Chinese approach to landscape. Both traditions are not mere imitations of nature but are capable of new juxtapositions creating a new sensation. We are especially inspired

24.12
CBD Beijing. By GROSS.MAX. / C.K. Yan, 2012.

24.13
Silica Valley. By GROSS.MAX. / Tom Chapman, 2017.

by the Chinese tradition of representing landscape as infinite scroll paintings. Like music or a film, and unlike western painting, the scroll follows the flow of nature's rhythms and is experienced as a journey in both space and time. For the central business district in Beijing, we designed a linear park with an artificial mountain range composed of red metal beams, constructed above an underground shopping canyon. As seen from different angles, the structure creates a different spatial experience. In a completely different reading, the mountain range can be interpreted as the cardiographic heartbeat or rise and fall of the stock exchange of the central business district.

Refer to Figure 24.13. A decade ago, Nobel-Prize-winning scientist Paul Crutzen suggested that we live in the new geological epoch of the Anthropocene. Humans have altered the planet sufficiently such that human interactions have reshaped landscape processes. As part of our proposal, for the transformation of the Sibelco Quarry in Heerlen, The Netherlands, we proposed a museum for the Anthropocene. The actual museum is situated as a contemporary cave dwelling in a relocated spoil heap mountain. Fossils found in the former coal mine were used to write the first flora of the Carboniferous era in the early twentieth century. Imprints of the fossils are displayed in the concrete walls. There is only one figure facing the viewer; by now, you may have recognized this man with a bowler hat leaning on his umbrella as he appears in nearly all of our images. For the connoisseur, it is a mark to verify the provenance of our images.

Notes

1. de Girardin, R.L. *De la Composition des Paysages*. Paris, France: Genève, 1777.
2. Gombrich, E. *Art and Illusion: A Study in the Psychology of Pictorial Representation*. New York, NY: Phaidon, 1960.
3. Ernst, M. "Beyond Painting." *Cahiers d'Art*, 1936, quoted in W. Spies (Ed.). *Max Ernst: Life and Work*. London, UK: Thames and Hudson, 2006, p. 55.
4. Breton, A. "Premiere exposition Dali." In: *Andre Breton: Oeuvres Completes: II*. Paris, France: Gallimard, 1992, pp. 307–309.
5. Bateille, G. "Le langage des fleurs." In: M. Bonnet (Ed.). *Documents*. Paris: Gallimard, 1929, 3(June).
6. Aphorism 280, titled "Architecture for the Perceptive," first published in: Nietzsche, F. *Die fröhliche Wissenschaft*. Chemnitz, Germany: Ernst Schmeitzner, 1882.

References

Bateille, G. "Le langage des fleurs." *Documents*, 1929, 3(June).
Breton, A. "Premiere Exposition Dali." In: M. Bonnet (Ed.). *Andre Breton: Oeuvres Completes: II*. Paris, France: Gallimard, 1992, pp. 307–309.
de Girardin, R.L. *De la Composition des Paysages*. Paris, France and Genève, Switzerland, 1777.
Gombrich, E. *Art and Illusion: A Study in the Psychology of Pictorial Representation*. New York, NY: Phaidon, 1960.
Nietzsche, F. *Die fröhliche Wissenschaft*. Chemnitz, Germany: Ernst Schmeitzner, 1882.
Spies, W. (ed.). *Max Ernst: Life and Work*. London, UK: Thames and Hudson, 2006.

Final Thoughts

Nadia Amoroso and Martin J. Holland

As demonstrated throughout this collection of essays, the professional commitment to express and incorporate useful and meaningful methods of communication and representation has long been established by the practitioners of landscape architecture. In the early twenty century, landscape architects investigated emerging advancements in communication and material to explore the potential benefits for the profession. As Halina Steiner has noted, Gertrude Jekyll's vast collection of writings and publications solidified her professional reputation as an 'artist-gardener.' Thaisa Way observes that it was Beatrix Farrand's eloquent facility in communication, ranging from conceptual sketches, to construction drawings, to engaged and lively personal correspondence with clients. Farrand's presence on site supervising projects under construction guided her projects to a successful resolution. Fletcher Steele's use of public lectures and the use of popular press assisted in establishing and refining his professional reputation among America's elite. Steele's speculative representational drawings were used to explore and test the aesthetic limits of his clients.

After the conclusion of the Second World War, a new spirit of innovation and potential emerged. Shannon Bassett describes how Isamu Noguchi blurred the conceptual—and disciplinary—lines between art and landscape through his dedicated use of clay study models as the material of professional inquiry, largely ignoring traditional drawing techniques in favour of three-dimensional sculpture as his preferred working methodology. For Roberto Burle Marx, Ana Rita Sa Carneiro describes his preoccupation with the primordial material of plants—the 'DNA of vegetation'—as the essential forms to be present within both his gardens and his paintings using the regional identity of his native Brazil as his inspiration. Jeffrey Blankenship details the significant influence that J.B. Jackson had upon the profession, not as a practitioner but as a gifted writer and cultural commentator, expanding the limits, values, and scope of the field of landscape architecture. Jackson brought new insight into the wider vernacular landscape primarily ignored as a topic of investigation. This conceptual expansion regarding the professional scope of practice led to advancements in the kinds of projects that landscape architects would engage in and led to innovation in how landscape architecture firms would self-organize, incorporate, and practice.

Kona Gray's piece on Edward D. Stone and EDSA details the necessity of consistency and clarity when describing visual communications, primarily when the firm is known for experimentation and innovation. As professional landscape architectural firms grew in size and influence, the need for a clear and constant ability to "express ideas through the simple medium of pen on paper" became paramount. Alison Hirsch's contribution regarding Lawrence Halprin identifies the city's urban condition as a renewed area of interest, with a recommitment to civic infrastructure and the public engagement processes required to allow for traditionally under-represented voices to be heard. In addition, Halprin developed a series of new drawing techniques (such as 'open scores' and 'motations') as additional representational methods to capture, augment, and test the vibrancy, spontaneity, and happenstance of city life. Similar freedom and sense of experimentation are revealed in Chip Sullivan's contribution admiring the works of two of his heroes, Garrett Eckbo and Thomas Church. Sullivan's work lovingly deconstructs the representative work of both landscape architects. It provides considerable technical details of the tools and the techniques they used within their design projects to express and capture the avant-garde spirit of post-war America. Frederick Steiner provides a detailed history of Ian McHarg's methodology that reveals the dynamic interplay between biophysical resources and cultural systems, ultimately resulting in the composite suitability map for which McHarg is famous.

As emerging practitioners in the 1980s and 1990s, Adriaan Gueze, Walter Hood, James Corner, and Ken Smith, like many of their contemporaries, utilized various methods of collage, composition, and repetition. This was a time of experimentation before the advancement of digital technologies and Photoshop. Embracing the Xerox black-and-white and colour copy machines, these landscape architects experimented with scale distortion, cutting and pasting, and the layering of copied elements.

Grounded in academia and practice, many of these firms explored ways to capture the viewers' attention by experimenting with collage as an artistic platform. They were not looking to portray the landscape designs 'as is,' but rather to capture the mood or atmosphere. For example, Yves Brunier would use 'crumbled' pieces of paper, coloured construction paper and other found objects like 'Q-Tips' as trees to quickly craft a Picasso-like study model. These crude models were child-like and whimsical, capturing the design intent and leaving the viewers curious about the final design. According to GROSS.MAX., the collage technique is a critical part of their design thinking, along with the presentation. GROSS.MAX. sees the collage/montage as 'assemblage of fragments and layers which at its best read as a palimpsest of time.' The crude-collage could be seen as means to represent indeterminate space, one that will change over time. These kinds of visualizations allowed the viewer to think about the landscape.

The introduction of Photoshop and other emerging digital technologies in landscape architectural offices in the 1990s and early 2000s drastically changed how many firms represented their ideas. Many firms continued to utilize collage techniques. Various vegetation, textures, the entourage of people, and other elements were scanned from magazines or photographs and then cut-and-pasted onto blank digital canvases and saved and stored into digital image banks. Once archived, these elements could then be easily accessed and layered into new landscape perspectives. The art scene in New York influenced Ken Smith in the early days of his career, and his graphic montage work reflected that art inspiration. In the late 1980s and early 1990s, montage representation was within the postmodern art context. He was inspired by landscape representation of Robert Smithson and his notion of Non-Sites.

In design competitions, firms may risk 'losing' if the images do not accurately represent the design. Leading on the safer side, firms may choose to develop hyper-realistic renderings that portray real environments. They may include the clear skies, lush greenery, 'happy' people, with the final touch of the photoshopped 'sun-burst' to make the rendering eye-catching. Making the rendering as realistic as possible leaves little room for interpretation. Experimental montages provide space for discussion, interpretation, and thought. It's an art form which could be risky for competitions, depending on the jury or the client. In the early days of digital representation, one could correctly guess the authors of the 'anonymous' design competitions by the visual communication styles and branding. Today, with the use of more advanced rendering engines and tools like Lumion, it is increasingly becoming difficult to distinguish who designed what since the visuals are becoming hyper-realistic and offering 'sameness' in terms of visual communication.

Hargreaves Jones and Gustafson Guthrie Nichol embraced clay as a medium of topographical expression. The models were not only valuable in understanding volume and landform, but they were also elegant pieces of art, which their audience could appreciate. For Balmori Associates, their renderings make the viewer more conscious about the space rather than the objects that occupy the space. Balmori developed a method of using a 'dot matrix' inspired by the halftone and the Ben Day process utilized by pop artist Roy Lichtenstein. Another of Balmori's methods for visualization includes using patterns rather than contours as a means to represent the spatial quality of their designs and the character of the landscapes. For Peter Walker, he is inspired by the minimalistic movement. He has also regarded landscape architecture as an art. He claims that landscape architecture is "an art that must be graded, built, and planted." This way of thinking is also apparent in his drawings and his teachings at Harvard's Graduate School of Design in the 1980s. He trained incoming non-design-educated students to create highly crafted models since many of these students could not draw like the BLA students who had more years of design and drawing training. Walker encouraged the students to make detailed models with properly scaled trees, proper topography, and realistic development of the ground plane. These models became a true representation of the space and allowed the viewer to see a 'real,' scaled-down version of the design in three-dimensional spaces. Many of these students became confident in their work due to this kind of visual representation. A number of these students later joined Walker's office, in which his firm often uses models to represent their designs.

Our hope as educators of landscape architecture is that this collection of essays centred around the topic of landscape representation helps students and practitioners alike to recognize the conceptual power of the medium of landscape. As indicated throughout this work, the ability to communicate innovative design ideas and the underlying theoretical principles that inspire the work is dependent upon the clarity, medium, and expression of the representation of the work itself. As unprecedented environmental and social change are spurred on by the global climate catastrophe, we expect that new and novel forms of representation will also emerge from the profession, offering informative and, perhaps, unexpected ways of addressing this existential threat.

Index

Note: **Bold** page numbers refer to tables. *Italic* page numbers refer to figures and page number followed by "n" refer to end notes.

Aalto, A. 68
abstracting 139–141
"Abstract World of the Hot- Rodder, The" (Jackson) 53, 55
Adams, T.: *Design of Residential Areas, The* 87
Adobe 113
Adobe Photoshop 61, 210
Against Nature (Huysmans) 204
Ai Wei Wei 126, *128*
algorithmic design 33
allegorical drawings 185–190
Allen, S. 163, *163*
Almost Cubist: A Tea Garden in Golden Gate Park, San Francisco 182, *183*
"Along the River during the Qingming Festival" (Zeduan) 90
American Academy, Rome 177–179
American Society of Landscape Architects (ASLA) 13, 24, 51
analogue-*versus*-digital conversation 61
Analysis Diagrams 59
Annuals & Amps: Biennials, the Best Annual and Biennial Plants and Their Uses in the Garden (Jekyll) 8
Antonioni, M. 200
Art Deco exposition (1925) 74
ASLA *see* American Society of Landscape Architects (ASLA)
ASPECT Studios 192–198; design representation 198; drawing 196–198, *197*, *198*; dynamic digital model 192–194, *193*, *194*; sketch diagram 194–196, *195*, *196*
AutoCAD 113, 125
Autoroutes du Sud project (Brunier) 118

Baldessari, J. 171
BAL/LAB 99, 112
Ballard, R. 200, 203
Balmori, D.: *Drawing and Reinventing Landscape* 101; *Landscape Manifesto* 100
Balmori Associates 99–101, 210
Bank of Early Bulbs, The (Jekyll) 9–10, *9*
Bardi, P. 47
Bartol, J.W. 25, 27, 30
Bassett, S. 209
Bass River Park 32
Bataille, G.: *Language of Flowers, The* 200–201
Baudrillard, J. 170–171
Beacon Equipment Company 29
Bealestreet flood *102*
Beaux-arts: beginnings of 67; mash-up 180–181
Ben Day process 210
Benjamin, W.: *One-Way Street* 205
Biblioteca degli Alberi *130*
Billerica Plan 86
BIM *see* Building Information Modelling (BIM)
Binney3 *102*
biomorphic heroes 67
Birth of Venus, The (Ernst) 117
Blaisse, P. 124–131
Bliss, A.B. 13; Rose Garden planting plan 15, *17*

Bliss, M. 15
Blockbuster Park *64*
Blues and Jazz Landscapes 186, *186*, 187
Bois de Boulogne park 133
Bois Vincennes park 133
Bonnard, P. 99
Borneo-Sporenburg, Amsterdam, the Netherlands Figure/Ground *134*
Boston Metropolitan Park 86
Botanical Garden of Dahlem, Berlin 44
Bowie, D. 203
Brambletye, lawn and herbaceous border at 7, *8*
Breton, A. 200
Breuer, M. 34, 72
BRIT *103*
Brooklyn Bridge Garden Mount *172*
Brown, L. "Capability" 116
Brunier, Y. 2, 125; *Autoroutes du Sud* project 118; collage as poetic procedure 117–118; displacement as figuration 117; figuring ground 118; greater fragmentation, fragment of 116–117; landscape representations 116–122; legacy 119; Museumpark *120–122*; plants, representation of 118–119
Building Information Modelling (BIM) 151
Burton Tremaine Residence, Santa Barbara, California *48*
By the Forces of Gravity (Ewan) 184n1
Byxbee Park, Palo Alto, CA 145

California Garden Movement 75
California Garden Series, The 74–75, *75*
California Garden Style 72
Camden High Line project 167, *168*
Campbell, J. 109
Candlestick Park, San Francisco, CA 145, *146*
Carmontelle, L. de 116
Cattleya Orchid and Three Hummingbirds (Heade) 200, *201*
CBD Beijing *207*
Charles Eliot, Landscape Architect (Eliot) 86–87
Chau, E.: *Random Imaginations* 184n1
Chevreul, M.-E. 4–5; colour wheel *5*; Law of Simultaneous Contrast 4; *Principles of Harmony and Contrast of Colour, The* 5
"Choose Your Own Adventure" 100, *104*
Cities (Halprin) 77
clay model *146*
clay site models 33
Clinton Presidential Center in Little Rock, AR 146
Cole Garden, Oakland, CA 1941 72, *73*
Coles, S. 196
collage, as poetic procedure 117–118
Collage, Ideogram, Greenport Harborfront, New York *2*
coloured pencil 61
"Colour in the Flower Garde" (Jekyll) 9
Commons and Campus Green, University of Cincinnati, OH 146
computers, introduction of 112–115, *113*
Concept Sketches 59

Confessions of an English Opium-Eater (Quincey) 200
connoisseurship, desire and cultivation of 24–30
Constant, J. 199
Construction Documents 59
Cook, P.: *Plug- In City* 1
Cool Grotto, A. 177, *178*
Corktown Commons 32
Corner, J. 1–2, 56; Camden High Line project 167, *168*; Collage, Ideogram, Greenport Harborfront, New York *2*; Downsview Park Competition, Toronto 163–165, *163–165*; eidetic drawings of 161–169; Field Operations 161, 163, *165*; field sketch *169*; High Line, New York City 161, 165–167, *166*; map-drawings 161; "Mind Landscapes: Navigation, Habitat and Imagination" 168; *Taking Measures Across the American Landscape* 162; windmill topography 161–162, *162*; 'Windmill Topography' 161
corporate drawings 107–108, *109*
Correio da Manhã 46
Crutzen, P. 207
cubism 72
cultural campus event fields 163
Cultural Landscape Foundation: *Laurie Olin on Design: Drawing is a Powerful Tool* 59–60
Cunha, E. da: *Os Sertões (The Backlands)* 44

Dawson Garden, Los Altos, California 1953–1955, preliminary planting study for 78
Delaware River Basin Study II *91*
"Descent into the Maelstrom" (Poe) 204
designer's preference 15
Design in the Little Garden (Steele) 25
Design Matters 59, 61–62
Design of Residential Areas, The (Adams) 87
Design with Nature (McHarg) 59, 91
De Smedt, J.: Maritime Youth House 32
Dion, M. 204
displacement as figuration 117
distorting 138–139
Domino Park, Brooklyn 161
Doncaster, England, regional plan of (1922) 87
Dooren, N. van 124
Downsview Park Competition, Toronto 163–165, *163*, *164*
Drawing and Reinventing Landscape (Balmori) 101
drawing experiments, for representing landscape 99–105; frame and peripheral vision 101; process 99–100; time 100–101
drawing "in" perspective 154–169
Drawing the Landscape (Sullivan) 184
Duany, A. 91
Dumbarton Oaks Garden 13–23; Arbor Terrace 16, *18*; "E" Terrace, planting plan for *22*; Eyrie Garden 15, *18*; Fountain Terrace *20*; garden terraces 15, *16*; Green Terrace to the Swimming Pool, circular stairs from *19*; Rose Garden 16, *17*, *19*; trace with graphite *21*
Dusseldorf, Germany, city plan for (1912) 87
dynamic digital model 192–194, *193*, *194*

Eagle 314 Draughting Pencil 68
Eastern Sea Storm Surge Barrier (Oosterschelde), Zeeland, the Netherlands *136*
Eckbo, G. 59, 70–76, 209; amoebas 72; birds-eye axonometric garden views 72; bombastic boomerangs 72; California Garden Series, The 74–75, *75*; Cole Garden, Oakland, CA 1941 72, *73*; cubism 72; Elementary School in Culver City, CA 1948 72, *73*; Estate in the Manner of Louis XIV 1934, An 70, *71*; Garden in Holmby Hills, Los Angeles, CA 1947 72, *74*; Garden in Woodside,

CA 1947 72, *73*; jazz 72; *Landscapes for Living* 67; Mineral King Co-operative Ranch, Community Park, CA 1939 74, *74*; modernism 72; Nickel Garden, Los Banos, CA 1944 *71*, 72; "quazy" atomic plants 72, 74–75; zig-zags 72
ecology 77–79
EDSA Graphics Book 59, 60
EDSA Style 59–66; evolution of 60–61; future of 61–62
Eidetic photomontage, Stockholm 166–167, *167*
Eight,' Bellevue, WA *152*
Elementary School in Culver City, CA 1948 72, *73*
Eliot, C. 86
Eliot, C.W.: *Charles Eliot, Landscape Architect* 86–87
Emerald Rules *171*
Ernst, M. 117–118, 200; *Birth of Venus, The* 117; *Garden of France, The* 117
Escher, M.C. 139
Escuintla master plan 65
Estate in the Manner of Louis XIV 1934, An 70, *71*
Ewan, R.F.: *By the Forces of Gravity* 184n1
Exposition Internationale des Arts Decoratifs Moderne (1925) 72

Fairchild Tropical Gardens 177
falls, Phoenix Lake, Marin County 1972 *78*
Farm Security Administration 57n8
Farrand, B.: Arbor Terrace 16, *18*; "E" Terrace, planting plan for *22*; Eyrie Garden 15, *18*; Garden for Witherell (Nathaniel), Greenwich, CT *14*, 15; landscape in prose and drawings, representing 13–23; Mr. Yew seat, construction drawing for *20*; Rose Garden planting plan 15, *17*; trace with graphite *21*
Festing, S. 10
Fifth Avenue *173*
figuring ground 118
"Fine Art in Landscape Architecture" (Steele) 24
Flight of the Spirits: Artists Inspired by Miyazaki 182
Foucault, M. 117
4-H club 170
fragmentation 116–117, 155
Frame 101
Freeways (Halprin) 77
French Ambassadors, The 156
Frensham Place, plan of *6*, 7
Friedrich, C.: *Wanderer Above the Sea Fog* 200
Frota, L.C. 47

Garden, The 9, 11
Garden and Climate 179, *180*
Garden and Forest 13
Garden Club of America, The 25
Garden for Witherell (Nathaniel), Greenwich, CT *14*, 15
Garden in Holmby Hills, Los Angeles, CA 1947 72, *74*
Garden in Woodside, CA 1947 72, *73*
Garden Magazine 25
Garden of Bacchus, The 180, *181*
Garden of France, The (Ernst) 117
Garden of the Phoenix, The 180–181, *181*
Garden of the Rose 181–182, *182*
garden reliquaries, and path to knowledge 181
garden representation 44–45
Gardens and People (Steele) 25
Garrison, W. 13
Geddes, P. 90, 91
geographic information systems (GIS) 88, 90
Geuze, A. 32
Gilgamesh Island 65
GIS *see* geographic information systems (GIS)

Gombrich, E.H. 200
Gomes-Martin, B. 184n1
Goods Line Project, The 196–198, *197*
Google's Flagship Headquarters Charleston East in Mountain View, CA 150
Governors Island, New York, USA *138*, *142*
Graduate School of Design (GSD), Harvard 170, 210
Grand Egyptian Museum, Cairo, Egypt 138, *139*
graphic communication 59–66
greater fragmentation, fragment of 116–117
Grosshopper® 193
GROSS.MAX 2, *3*, 210; AIR Rotterdam *201*; CBD Beijing *207*; Garden for a Plant Collector 203–204, *203*; Gleisdreieck *205*; Global Warming/Local Freezing *203*; imagery, iconograhy of 199–207; New Town/ New Nature 202–203, *202*; Old Town New Town No Town *205*; Rottenrow Garden *202*; Silica Valley *207*; Tempelhof Airport *206*; vertical garden *204*; Xi'an Master Garden *206*
growth of representation 107–115
Grubbs, C. 108
GSD *see* Graduate School of Design (GSD), Harvard
Gustafson, K. 1, 154–157, *157*
Gustafson Guthrie Nichol (GGN Ltd) 154, 156, 157, 210; 3D print of water feature with pencil annotations *160*; Hazelwood Green *158*, *159*

Hair Gardens Type One, Bouffant Topiary *174*
Halley, P. 171
Halprin, A. 79; "Happenings" 79, *80*
Halprin, L. 59, 77–85, 107, *108*; choreography 79; *Cities* 77; Dawson Garden, Los Altos, California 1953–1955, preliminary planting study for *78*; ecology 77–79; falls, Phoenix Lake, Marin County 1972 *78*; *Freeways* 77; High Sierra sketches *79*; Motation 79, *81*, *82*, 83, 85n4; *RSVP Cycles, The* 77; *Taking Part* 77; water sketches 83–85, *83*, *84*
Hamilton, R. 200
Hammer Museum, Los Angeles *128*
Hammock, T. 184n1
Hanging Gardens of Queens *175*
"Happenings" (Halprin) 79
Hargreaves, G. 1
Hargreaves Jones Landscape Architecture (formerly Hargreaves Associates) 144–153; client & community engagement 147, *148*, *149*; conceptual design 145–147, *146*; design, refining 148, 150–151; landscape, post-occupancy 151–153; site acquaintance 144–145, *145*
Harris, D. 185
Harrisburg *101*
Harvard Graduate School of Design 72, 107
Hazelwood Green *158*, *159*
Heade, M.J.: *Cattleya Orchid and Three Hummingbirds* 200, *201*
Herzog, W. 200
Hickingbotham Garden, Hillsborough, CA 1948 70, *70*
High Line, New York City 161, 165–166, *166*, 186
High Sierra sketches *79*
Hills, G.A. 87, 88
Himmel Uber Berlin (Wenders) 205
Hockney, D. 125
Holbein, H. 156
Holl, S. 185
Hood, W. 1; *Urban Diaries* 1
Hoover, H. 25
Horticulture 25
Hotel St. James 119
House Beautiful 25

Houston Botanic Garden, Texas, USA *143*
Hubbard, H.: *Introduction to the Study of Landscape Architecture, An* 179
Hutton, J. 203
Huysmans, J.K.: *Against Nature* 204
Hyper Lane development, Chengdu, Sichuan 194

Ian Potter Children's Wild Play Garden 198, *198*
Ibirapuera Park, Sao Paulo: geometric garden in *49*; plan of *48*
Impulse 77
Ingels, B.: Maritime Youth House 32
Introduction to the Study of Landscape Architecture, An (Hubbard and Kimball) 179

Jackson, J.B. 51–57, 209; "Abstract World of the Hot-Rodder, The" 53, 55; brown paper *52*; landscape, illustrating 52–56; "Need of Being Versed in Country Things, The" 56; personal reconnaissance 51–52; "Stranger's Path, The" 53, *54*; symbols, usage of *53*; *Taking Measure Across the American Landscape* 56–57; "To Pity the Plumage and Forget the Dying Bird" 55
Jacobs, J. 133
Jardins de Metis International Garden Festival 100, 101
jazz 72
Jekyll, G. 4–11, 22, 209; *Annuals & Amps: Biennials, the Best Annual and Biennial Plants and Their Uses in the Garden* 8; *Bank of Early Bulbs, The 9*, 9–10; Brambletye, lawn and herbaceous border at 7, *8*; collecting skills 4–7; "Colour in the Flower Garde" 9; Frensham Place, plan of 6, 6–7; Munstead, herbaceous border at *10*; West Dean Park, herbaceous border plan for 7, *7*; writing 8–11
JJR *see* Johnson, Johnson, and Roy (JJR)
Jobs, S. 95, 112
John F. Kennedy Performing Arts Center *62*
Johnson, B. 108, 115n3
Johnson, Johnson, and Roy (JJR) 115n3
Jones, B.C. *see* Farrand, B.
Joyous Science, The (Nietzsche) 206

Kahn, L. 76
Kandinsky, V. 156
Karson, R.S. 24, 25
Kaueper, G. 59
Ken Smith Workshop 171, *173*
Kerouac, J. 74
Kimball, T.: *Introduction to the Study of Landscape Architecture, An* 179
Koolhaas, R. 117
Kruger, B. 171

Laban, R. 85n4
Ladies Home Journal 25
Landscape 51–53, 56, 57, 57n8; back cover of *54*; front cover of *54*, *55*
Landscape Architecture 25
Landscape Architecture Quarterly 51
Landscape Manifesto (Balomi) 100
Landscapes for Living (Eckbo) 67
Language of Flowers, The (Bataille) 200–201
Laurie Olin on Design: Drawing is a Powerful Tool 59–60
Law of Simultaneous Contrast 4
Lawrence Halprin & Associates 77
layer-cake diagram 89
layered approach 136, *137*
Le Corbusier 68, 204
LEED system 152
Lerup, L. 185

Levine, S. 171
Lewis, P. 87, 88
Libeskind, D. 1
Lichtenstein, R. 99, 210
Lincoln Centre, New York City 32
Lister, N.-M. 165
London County plan (1943) 87
Lorrain, C. 199
Louisville Waterfront Park, KY 146, *148*, 150
Lumion 100
Lutyens, E. 6

Mack, M. 185, 186
MacKaye, B. 91
MacLean, A. 56
Madrid Nuevo Norte *103*, *104*
Malevich, K. 156
Manfredi, W.: Seattle Sculpture Park 32
Manning, W.H. 24–25; Billerica Plan 86
maps, as slices of time 88–90, *90–93*
Martin Garden, Aptos, CA 1947, Preliminary Plan for 68, *69*
Marx, R.B. 43–49, 209; Burton Tremaine Residence, Santa Barbara, California *48*; garden representation 44–45; Ibirapuera Park, Sao Paulo *48*, *49*; Ministry of Education and Health, roof garden of *47*; mystery in interior of matter 45–49; *Praça de Casa Forte* 44, *45*; *Praça do Ministério do Exército* 47; *Praça Euclides da Cunha* 44, *46*
Maya® 193
McHarg, I.L. 59, 209; complex process, mapping 86–95; layer-cake diagram *89*; maps, as slices of time 88–90, *90–93*; suitability analysis 91–94, *93*, **94**
M'Closkey, K. 162
Medford Township 91, *94*
Metis CYOA Time-Process *104*
"Mind Landscapes: Navigation, Habitat and Imagination" (Corner) 168
Mineral King Co-operative Ranch, Community Park, CA 1939 74, *74*
MiniCAD 113
Ministry of Education and Health 45; roof garden *47*
Miro 100
Mitchell, W.J.T. 101, 186
Miyazaki, H. 182
modernism 72
Moerenuma Park, Japan 35, *41*
Mohammed Bin Zayed City Master plan *64*
Moore, H. 76
Morris, W 5
Moses, R. 35
Motation 79, *81*, *82*, 83, 85n4
Mr. & Mrs. Charles Shuey, Claremont, CA 1937, Preliminary Plan for 67–68, *68*
multiplicity 155
Municipal Art Society *173*
municipal landscape competition *130*
Munstead, herbaceous border at *10*
Museum of Modern Art, New York 47
Museumpark *120–122*
My Neighbour Totoro 182

Nagy, M. 204
National September 11 Memorial 108
"Need of Being Versed in Country Things, The" (Jackson) 56
Nelson, G. 87
New York, regional plan of (1929) 87

New York Times 10
Nickel Garden, Los Banos, CA 1944 *71*, 72
Nielsen, A. 124
Nietzsche, F.: *Joyous Science, The* 206
Noguchi, I. 107, 209; Moerenuma Park, Japan 35, *41*; playgrounds 34–35; Play Mountain 35, *40*; Reader's Digest Building, Tokyo 34; Riverside Park Playground, New York City 33, 35, *38–40*; sculptured landscapes 32–41; UNESCO Garden 34, *36*, *37*
Non-Sites 170–175
Northside Park in Hunters Point, San Francisco, CA 147, *149*
Nouvel, J. 118, 119

Olin, L. 60, 94
Olmsed's big North American park 133
Olmsted, F.L. 14
One-Way Street (Benjamin) 205
Ontario Province, Hills' plan for 87
Os Sertões (The Backlands)(Cunha) 44
Oudolf, P. 125

Pamphlet Architecture 185
Pantone 87
parametric modelling 33
Parral, J. 184n1
Parsons, S. 13
Patterson, R. 14, 22
Pelli, D. 101
Penn's Landing, Philadelphia, PA *152*
Pergola of Maximapark, Utrecht, the Netherlands *141*
peripheral vision 101
Perth's Yagan Square 193
PESTEL 133
Peter Walker and Partners (PWP) 112, 115
plants, representation of 118–119
Plater-Zyberk, E. 91
Play Mountain 35, *40*
Plug-In City (Cook) 1
Poe, E.A.: "Descent into the Maelstrom" 204
Pont Royal 63
Portable Garden Stourhead *171*
Porter, N. *157*
Post-occupancy, Crissy Field, CA *153*
Power-Points 109
Praça de Casa Forte 44, *45*
Praça do Ministério do Exército 47
Praça Euclides da Cunha 44, *46*
Prince, R. 171
Principles of Harmony and Contrast of Colour, The (Chevreul) 5
"Programming the Urban Surface" (Wall) 33–34
Progressive Architecture 85n7
Propes, E. 61–62
PWP *see* Peter Walker and Partners (PWP)

Queen Elizabeth Olympic Park, London, UK 144, 148, *150*
Quincey, T. de: *Confessions of an English Opium-Eater* 200

Random Imaginations (Chau) 184n1
Reader's Digest Building, Tokyo 34
Reed, C. 32
Revit® *152*, 193
Rhinoceros® (Rhino3D® digital modelling software) 157
Rhino® 100, 193, *194*
Rio Grande Valleye, New Mexico 56
Riverside Park Playground, New York City 33, 35, *38–40*

Robinson, W. 6
Rockefeller, A.A. 15, *18*
Rose, G. 8
Roshen Chocolate Factory *129*
Rotonda's Orientation, The 178–179, *179*
Royal des Gobelins 5
RSVP Cycles, The (Halprin) 77

sacred woodland spirits 182–184
Saenredam, P. 199
Sargent, C.S. 13
Sasaki, H. 59
Sasaki Associates 177
Sasaki Walker and Associates (SWA) 108, *110*
Schouwburgplein, Rotterdam, the Netherlands *137*
Scissortail Park 147
Scofidio, D.R. 32
Seattle Center Fountain *84*
Seattle Sculpture Park 32
Seminary South Shopping Center Fountain *84*
Shelter from the Wind 177, *178*
Shen, S. 184n1
Sibelco Quarry, Heerlen, The Netherlands 207
Silica Valley *207*
Silver Lake Reservoir Complex Master Plan, Los Angeles *145*
Silver Lake Reservoir Master Plan, Los Angeles, CA 147, *149*
simulacra 171
simulation 171
SITES system 152
SketchUp 113
Smith, W. 91
Smithson, R. 170
Soil Conservation Service 56, 57n8
soundwaves: growth *105*; rollview *105*
splicing 61
Standard Oil Company 57n8
Stedelijk Museum Extension competition 126, *126, 127, 129*
Steele, F.: connoisseurship, desire and cultivation of 24–30; *Design in the Little Garden* 25; "Fine Art in Landscape Architecture" 24; *Gardens and People* 25; as member of The Garden Club of America 25; swimming pool design 25–30 *26–30*
Steiner, H. 209
Stone, E. *62*, 209
STOSS 32
Stout, B. 185
"Stranger's Path, The" (Jackson) 53, *54*
Streatfield, D. 75
Student Union Plaza, University of California, Berkeley: motation study for *81*
suitability analysis 91–94, *93*, **94**
Sullivan, C. 209; Almost Cubist: A Tea Garden in Golden Gate Park, San Francisco 182, *183*; Cool Grotto, A 177, *178*; *Drawing the Landscape* 184; Garden of Bacchus, The 180, *181*; Garden of the Phoenix, The 180–181, *181*; Garden of the Rose 181–182, *182*; Rotonda's Orientation, The 178–179, *179*; Shelter from the Wind 177, *178*; Totoro's Sacred Tree 182–184, *183*; Villa Rotunda, Vicenza, Italy 178, *179*
Sunrise Saudi Arabia *63*
sustainable landscapes, design vocabulary for 179–180
SWA *see* Sasaki Walker and Associates (SWA)
swimming pool design 25–30, *26–30*
system of expression 4–11

Taking Measure Across the American Landscape 56–57, 162
Taking Part (Halprin) 77
Tarkovsky, A. 200
Third Street Light Rail, San Francisco *175*
Tommy (Thomas) Church 74–76, 209; artistry 67–68; elegance 67–68; Hickingbotham Garden, Hillsborough, CA 1948 70, *70*; Martin Garden, Aptos, CA 1947, Preliminary Plan for 68, *69*; Mr. & Mrs. Charles Shuey, Claremont, CA 1937, Preliminary Plan for 67–68, *68*; Zelinsky Garden, Atherton, CA 1947 68–70, *69*
Tongva Park, Los Angeles 161
"To Pity the Plumage and Forget the Dying Bird" (Jackson) 55
Toronto Waterfront, Lake Ontario, Canada *140*
Town and Country Planning Textbook (Tyrwhitt) 87
Transforming the American Garden: 12 New Landscape Designs 181
Treib, M. 33
Turner, J.M.W. 4
Tyrwhitt, J. 91; *Town and Country Planning Textbook* 87

Underline, Miami 161
UNESCO Garden 34, *36, 37*
UNESCO House 34
University of Pennsylvania School of Design 161
Urban Diaries 187–190, *187–190*
Urban Diaries (Hood) 1
Urban Gallery, The 194–195, *195, 196*
"Urbanography" 85n7
USGBC: LEED and SITES systems 152

Valkenburg, M. van 181
Valkenburgh, M.V. 7, 32
Valladares, C.do P. 43, 47
van der Rohe, M. 204
Vicobello, Siena, Italy, sixteenth century 179, *180*
Victorian Desalination Plant and Ecological Reserve 193, *193*
Village of Yorkville Park Concept Montage *173*
Villa Rotunda, Vicenza, Italy 178, *179*
Villa Torlonia, Frascati, Italy 179–180, *180*
Villa Vizcaya 177
von Humboldt, A. 91

Walker, C. 113
Walker, D. 112
Walker, P. 107–115, 210; artist drawing for character 108, *111*; computers, introduction of 112–115, *113*; corporate drawings 107–108, *109*; digital drawings *114*; early landscape drawings 107, *108*; exceptional drawings 108, *110*; models empower the curriculum 109, 112; photography to illustrate character 109; plant materials 109; scale 109; video and movement 115
Walker Art Center *128*
Wall, A.: "Programming the Urban Surface" 33–34
Wanderer Above the Sea Fog (Friedrich) 200
Warhol, A. 200
water 83–85, *83, 84*
watercolour 14
weekend watercolour excursions 177
Wenders, W. 200; *Himmel Uber Berlin* 205
West 8: abstracting 139–141; alignment of 133–136; constructed landscapes, legacy of 132; distorting 138–139; hand of the painter 141–143; innocent landscape 133; layered approach 136, *137*; strategic editing 137; visual communication of 132–143
Westward Ho 67

White, S. 115n1
William J. Clinton Presidential Center, Little Rock, AR *148*, 150
Williams College 24
windmill topography 161–162, *162*
Woods, L. 1

Xerox Photocopy Machine 134
Xi'an Master Garden *206*

Yale University: Sterling Memorial Library 100
Yongsan Park, Seoul, South Korea *140*, *141*
Yongsan Park project, Korea 138
Your Private World 70

Zeduan, Z.: "Along the River during the Qingming Festival" 90
Zelinsky Garden, Atherton, CA 1947 68–70, *69*
Zip-A-Tone 87, 107, 115n2, 170